Hog Wash

Hog Wash

The Strange Case of SSA Global

John W. Sutherlin

HAMILTON BOOKS
AN IMPRINT OF
ROWMAN & LITTLEFIELD
Lanham • Boulder • New York • London

Published by Hamilton Books
An imprint of The Rowman & Littlefield Publishing Group, Inc.
4501 Forbes Boulevard, Suite 200, Lanham, Maryland 20706
www.rowman.com

86-90 Paul Street, London EC2A 4NE, United Kingdom

British Library Cataloguing in Publication Information Available

Library of Congress Cataloging-in-Publication Data

Names: Sutherlin, John W., author. Title: Hog wash : the strange case of SSA Global /
John W. Sutherlin.
Description: Lanham, Maryland : Hamilton Books, [2024] | Includes bibliographical
    references and index. | Summary: "Hog Wash documents the decade-long effort of
    a group of international collaborators to 'solve' the waste problem from industrial
    animal farming which poisons lakes, rivers, and streams"—Provided by publisher.
Identifiers: LCCN 2023045078 (print) | LCCN 2023045079 (ebook) | ISBN
    9780761874188 (paperback) | ISBN 9780761874195 (epub)
Subjects: LCSH: Entrepreneurship. | Factory farms—Environmental aspects. | Miller-
    Kurakin, G. B. | Sutherlin, John W.
Classification: LCC HD62.5 .S8893 2024 (print) | LCC HD62.5 (ebook) | DDC
    658.1/1—dc23/eng/20231030
LC record available at https://lccn.loc.gov/2023045078
LC ebook record available at https://lccn.loc.gov/2023045079

∞™ The paper used in this publication meets the minimum requirements of American
National Standard for Information Sciences—Permanence of Paper for Printed Library
Materials, ANSI/NISO Z39.48-1992.

Contents

A Note from the Author

In order to protect privacy, the names and identifying details of the individuals in this book have been changed.

Preface

An apology is warranted.

It takes me several chapters to get to the heart of this saga. So, I am sorry. Be patient.

Yet, in many ways, the end was foretold. It was inevitable. I just did not know it.

By the time it ended, like many endings, it seemed to happen rather swiftly.

On October 23, 2009, George Kurakin-Miller died. He was 54 years old. He was a man of the world. Educated. Cosmopolitan. Incredibly sophisticated.

George had a huge smile and a bigger heart. He loved his family, especially his mother. Perhaps the shared bond of having witnessed his father being gunned down in front of his mother forged an unshakeable union. People exposed to tragedy early respond in so many different ways. Some people never quite recover. He was different. George stepped into the role of *Head of Household* for the Kurakin-Miller Family. Everyone depended on George. He became the 'man of the house' and shouldered more than anyone else. This allowed his family to pursue their dreams while he educated himself and worked hard.

His demise signaled the end to more than a decade of international meetings, high-profile presentations, back-breaking research on two continents, and millions of dollars in investments. George's death was the final nail in a coffin lined with the dreams of dozens of people from dozens of countries. Many of these people were honorable, decent entrepreneurs, government officials, and university researchers (like me and my partner Juan).

Others were not quite right. Dodgy suspects with insane dreams of wealth. They were a nefarious assortment of international lawyers and financial consultants with ill-defined nationalities and unproven moral convictions. Mostly rascals with little more than the desire for money. Yet, George held this group together.

Now he was gone.

The PROJECT quietly and quickly fell apart. My research partner had left for California, too. Failure was always the most likely outcome of any such venture as ours. Most such international efforts end without 100 percent of whatever is being proposed actually coming to fruition. It might be fair to say that even before George's body was laid to rest, his ambitious and courageous efforts were doomed. Only in hindsight does it seem so obvious. This project was never going to flourish. George's dream and our hard work were not enough to overcome the paradigm of pollution entrenched in a global industry unable and unwilling to change.

Our goal: clean-up streams, rivers, and lakes polluted with the vile waters discharged from confined animal feeding operations (CAFOs). Mainly from industrial hog farms. Millions of gallons of unregulated wastewaters fouled the air and water with the stench of pig feces and urine.[1] For more than 60 years, researchers have pushed the idea that large lagoons with sufficient depth could sequester odors and pollution.[2] The large volume of waters are made worse by the concentrations of ammonia (a hazardous substance) present in the waste stream.[3] More current research has shown nitrous oxide emissions coming from these lagoons.[4]

Some research suggests that ammonia and nitrous oxide from pig farms is one of the largest sources of air and water pollution.[5] Further, even if lagoon "technology" was a sound solution for hog slurry, nasty weather during, for example, hurricanes causes these small ponds full of manure, urine, and water to become overwhelmed and spill into receiving waters without any anaerobic treatment or volume reduction.[6] In addition, "When the lagoons become full, the waste water is often sprayed onto fields as nutrients for crops. The waste, which contains harmful bacteria like E. coli or salmonella, can wash off into local waterways and cause groundwater contamination and fish kills."[7]

In 1999, this problem came to a head in North Carolina following Hurricane Floyd. People around the world took notice. To be fair, I was largely unaware of the scope of the problem worldwide. I was close to completing my PhD while working at an Environmental Protection Agency (EPA) funded think-tank and policy center at the University of New Orleans. I was also teaching across town at Tulane University. About to get married. Maybe even start a family. Wrapping up a documentary film series and about to start another one. I had no idea what I was about to get sucked into. Like falling head-first into the thick muck of a dirty lagoon, I was about to get baptized in hog wash.

Literally.

The environmental catastrophe of industrial hog farming was about to push all my other aspirations aside. After all, I was about to become part of an international group of problem-solving experts poised to make huge amounts of money. Pig waste was not just a local issue. It was becoming a global concern.[8] People love to eat pork and that means lots of CAFOs everywhere, which spews that distinctive aroma of pig poop into the air while torrential hog water slurry dumps into fresh water. Across Europe, especially the northern countries,[9] and the UK, environmental and animal organizations, regulatory and other government agencies, and agricultural universities were looking for options.[10]

It is also important to note that pig waste is but one of the toxic wastes at industrial hog farms. One of the nation's first Superfund Sites was a pig farm. According to the Environmental Protection Agency (EPA), the 8-acre Picillo pig farm (Rhode Island) was contaminated with hog waste and "[m]ore than 10,000 drums of hazardous waste and an undetermined bulk volume of liquid chemicals were disposed of into several unlined trenches."[11] Here, state and federal actions seemed inadequate to address the immense ecological catastrophe.[12] No one seemed to have a viable solution. No one had an environmental alternative that was cost-effective. Even when addressing only the hog waste, there seemed to be no answer. My research partner and I thought we had something environmentally friends and economically attractive. Looking back, I cannot say if we ever had anything fitting either of these conditions. Maybe. It would have been nice to find out, though.

As a note, this story comes with a warning. Much of this is taken from my experiences and recollections. In other words, I am certain that my own biases have entered into this work. I am also certain that I have made mistakes or just plain forgot what I once thought I knew. Those mistakes are not deliberate. I know that I made many mistakes then and now and am responsible for so much that went wrong. But, over time, as I reminisce and ponder the past, I have more than likely mentally assigned blame elsewhere. I doubt that I am even conscious of doing that. There is no attempt here to claim innocence. There were people that I never liked that now appear to be much better. There are others that I should never have ever liked or trusted. They probably feel the same about me and they are probably right.

Here, though, I am laying out the "facts" as I experienced them first hand. This is neither an apology nor a confession. This is the tale of a group of people that tried very hard to reduce or eliminate a serious environmental problem. Many of the names and identifying details of people and events have been changed to protect their privacy and the privacy of their families. But they all were very real and we led by George Kurakin-Miller.

Despite the hog wash, there was some really good work here. Nuggets of truth, real science. I continue to anticipate and react to dubious ventures. The result often leads others to think that I am negative, or hardheaded, or constantly looking for failure. They could not be more wrong. Toughness and durability are never liabilities. Ever project requires a truth-teller.

My opinions of others here should not be interpreted as handed down to Moses after the Ten Commandments. There are certainly errors made in developing this book. None were deliberate. Whatever pearls once existed have surely been trampled by the swine. But there were great times as well.

I think. Memories are funny things. Very few things are as we remember them.

There will be many readers that will be skeptical of the events, or, worse, critical of my role. I have heard many times over the past decade how shockingly naïve I must have been to fall for such outlandish nonsense. To be fair, most people fall prey to their own pride and greed.[13] I was no different. Vanity has to be worst sin. But, hopefully, that was a different time.

This book attempts to capture the spirit of that era.

This is the strange case of SSA Global.[14]

I am, of course, grateful to my family for their patience and support during these years.

John W. Sutherlin, PhD
July 2023

NOTES

1. Clark, C. E. (1965). Hog waste disposal by lagooning. *Journal of the Sanitary Engineering Division, 91*(6), 27–42.

2. Hart, S. A., & Turner, M. E. (1965). Lagoons for livestock manure. *Journal (Water Pollution Control Federation)*, 1578–1596.

3. Stoltenberg, D. H., & McKinney, R. E. (1966). Discussion of "hog waste disposal by lagooning." *Journal of the Sanitary Engineering Division, 92*(4), 78–81.

4. Mkhabela, M. S., Gordon, R., Burton, D., Smith, E., & Madani, A. (2009). The impact of management practices and meteorological conditions on ammonia and nitrous oxide emissions following application of hog slurry to forage grass in Nova Scotia. *Agriculture, Ecosystems & Environment, 130*(1–2), 41–49.

5. Anderson, N., Strader, R., & Davidson, C. (2003). Airborne reduced nitrogen: ammonia emissions from agriculture and other sources. *Environment International, 29*(2–3), 277–286. Pigs are not the only ecological culprit here. Chicken waste are also particularly noxious. A good source here is the following. Magbanua Jr, B. S., Adams, T. T., & Johnston, P. (2001). Anaerobic codigestion of hog and poultry waste. *Bioresource Technology, 76*(2), 165–168.

6. Oglesby, Cameron. Environmental health news. Hurricane season spurs hog waste worries in North Carolina. Located at https://www.ehn.org/north-carolina -hurricanes-hog-farms-2652972415/hog-farms-in-north-carolinas-100-year-flood -plain. Accessed January 13, 2022.

7. Ibid.

8. Webb, J., Menzi, H., Pain, B. F., Misselbrook, T. H., Dämmgen, U., Hendriks, H., & Döhler, H. (2005). Managing ammonia emissions from livestock production in Europe. *Environmental Pollution, 135*(3), 399–406.

9. Koerkamp, P. G., Metz, J. H. M., Uenk, G. H., Phillips, V. R., Holden, M. R., Sneath, R. W., . . . & Wathes, C. M. (1998). Concentrations and emissions of ammonia in livestock buildings in Northern Europe. *Journal of Agricultural Engineering Research, 70*(1), 79–95.

10. Webb, J., & Misselbrook, T. H. (2004). A mass-flow model of ammonia emissions from UK livestock production. *Atmospheric Environment, 38*(14), 2163–2176.

11. Environmental Protection Agency. Superfund site: Picillo Farm Coventry, RI. Cleanup activities. Located at https://cumulis.epa.gov/supercpad/SiteProfiles/index .cfm?fuseaction=second.Cleanup&id=0101284#bkground. Accessed 15 May 2023.

12. Muller, B. W., Brodd, A. R., & Leo, J. (1982). Picillo Farm, Coventry, Rhode Island: A superfund & state fund cleanup case history. In *Proceedings of the National Conference on Management of Uncontrolled Hazardous Waste Sites*, November (pp. 268–273). By the time the site was listed as a Superfund, containing all of the toxins was a major issue. Boving, T. B., & Blue, J. (2002). Long-term contaminant trends at the Picillo Farm superfund site in Rhode Island. *Remediation Journal, 12*(2), 117–128.

13. Konnikova, M. (2017). *The confidence game: Why we fall for it . . . Every time.* Penguin.

14. There were various companies that were organized during this project. One was International Waste Management Systems. Another was Basalt Limited. There were several more. Companies House (UK.gov) has a full listing of all of these.

Chapter 1

What a Terrible Idea!

I grew up in the country. Deep in the woods. Even then, I knew early on that country people do not live near people for many reasons. One of those reasons was they hate government. Farmers are self-sufficient. People who live on farms do not advocate for more government involvement into their personal lives or business affairs. That is for city-folks. Also, it is harder to regulate decentralized populations. As such, historically, agricultural operations have been poorly or unregulated industries for decades.[1] If they wanted bureaucratic intrusion, they would move to town. Of course, farmers want electricity and water, but little else.[2] Yet, when farm activities impact the ecology and public health of everyone, urban, suburban, and rural, then government must step in. Despite often hostile views towards government "intrusion," farmers fund they must cooperate with federal regulators.[3] As such, large-scale animal farming demands government oversight.

In order to "put meat on the table" American farmers began centralizing animal production in the 1950s to reduce costs of labor, management, and transportation of (not so little) "piggies to market." The industrial farm animal production (IFAP) model relies upon massive farms with animals stuffed into tiny cages often piled on top of one another, screaming.[4] There can be "cities of pigs" numbering in the hundreds of thousands. I visited one with almost 500,000 pigs in North Carolina where waste was drained into adjacent pits. That seems to be all too common. It also seems to be unsustainable.[5] Yet, we continue with this insufferable system.

According to the US Government Accountability Office (GAO), between 1982 and 2002, the number of concentrated animal feeding operations (CAFOs) increased by 246%.[6] Demand for pork seems to be insatiable. Large-scale, corporate livestock production facilities, require, *inter alia*, massive waste management efforts.[7] The reason should be obvious: waste from industrial pig operations is massive. Depending on the type of facility, wash-down frequency and volume, and dietary demands (which influences urine

1

and feces production and nutrient content), a single pig can generate almost 3 gallons of liquid and 11 pounds of manure . . . daily![8]

When all of the cows, chickens, and pigs are included, "There are approximately 450,000 CAFOs currently operating in the USA, with the majority located in watersheds feeding major riverine and estuarine systems with known water quality problems."[9] In the CAFOs in North Carolina, with more than 4 million pigs, that is a river of pee and a mountain of manure.[10] Unlike smaller, family owned farms where waste (mainly manure and urine) can be land-applied and used as soil amendment or compost,[11] waste from CAFOs drain into open pit lagoons (see Figure 1.1 on the following page; notice the thick layer of manure on the top of the liquid portion). Surface water quality in North Carolina has suffered because of CAFOs.[12]

This transformation has been critical for reducing the price of pork, but it has come with a huge environmental cost.[13] CAFOs create intense environmental and worker safety concerns, for example, due to excessive amounts of ammonia, NH_3, (and other toxins) in the air and water.[14] Other research has further documented other harmful metals bioaccumulating in the soil around CAFOs.[15] It should not be ignored that the CAFO system results in a miserable life for the pigs as well.[16]

Being a pig sucks.

Adjacent water quality suffers as CAFOs drain urine and manure into poorly designed lagoons. The solids sink to the bottom to rot over time. This

Figure 1.1.   Lagoon where most pig waste is drained. Courtesy of the Author.

is called anaerobic digestion. As the solids decompose, the amino acids in the waste breakdown and produce ammonia vapors and methane gas.[17] The smell is unbearable. But it is what you do not smell that can really hurt you. Washdown water and urine are the main sources of liquid wastes. This fraction of the waste drains or is pumped to an adjacent lagoon for storage and to cover the solids.

The liquids inevitably leak or overflow following heavy rain.[18] The ground where the lagoon overflows often looks like someone burned the ground because the nutrients and contaminants have overloaded nature. Besides the ammonia and metals found in the lagoon waters, hormones and antibiotics that pass through the pig's digestive system wash into rivers, streams, and creeks.[19] There have been some efforts to reduce the cause of excess ammonia emissions and discharges by regulating the nitrogen in the diet of pigs.[20] But the digestive system of a pig is very different than, say, cows or chickens, and must rely on grains.[21] Also, early studies found that pigs do not digest food the same way other animals do and are very inefficient. That means that farmers must "overfeed" pigs resulting in much more waste per pound of "finished" pig when it goes to market.[22] The most recent environmental monitoring and analysis (2021) revealed that the problem of too much waste with too many nutrients being undigested remains.[23] Regulatory efforts, which are "inconsistent" at best, have largely been somewhere between grossly inadequate and a dismal failure.[24]

So, pigs cramped in cages, stacked on top of one another, filled with pharmaceuticals produce loads of waste that are "treated" in leaky lagoons so that pork-eaters have cheap bacon and pork chops? And yeah, I know that "bacon tastes good, and pork chops taste good," but at what cost to the planet?[25]

Until I started working on this project, I was a life-long, card-carrying member of the Hog Eater Club (not a real organization as far as I know). The ecological burden coupled with the inhumane conditions for pigs and the terrible stench compelled me to find other culinary choices.

There are essentially two major sources of pollution from CAFOs: water and air. Both are regulated by federal, state, and local governments. On a federal level, the *Clean Water Act* (CWA) required periodic expansion of industries obligated to get a National Pollutant Discharge Elimination System (NPDES) permit. NPDES permits are issued and enforced by state governments. The promulgation of regulations by the federal government is essential for most laws. But many environmental groups felt like the Environmental Protection Agency (EPA) had failed to revise the CWA and industries, such as CAFOs, were operating (and profiting) essentially unregulated. Cities, power plants, and refineries had to comply with the CWA. The local car wash is required to meet certain discharge limits. So why not multi-million-dollar factory-farms? The Natural Resources Defense Council took the EPA to court and more than a dozen new industries were added to the list.[26]

Water is the first, most obvious environmental issue. Air emissions from CAFOs are a major concern as well. For more than 20 years, the EPA, as part of their mandate from the *Clean Air Act* (CAA), and the US Department of Agriculture (USDA) through a joint task force Board on Agriculture and Natural Resources (BANR) have evaluated data, technologies, and policies and yet there is little to show except public outrage.[27] But not enough outrage where consumers are willing to pay more for pork chops and bacon. Public outrage often manifests itself in the form of odor complaints. The smell from CAFOs and the lagoons is horrible and destroys quality of life.[28] Pig waste is distinctive and unpleasant. Unlike, say, cow or horse manure, which has a mildly sweet aroma, I have never heard anyone say, "I love the smell of pig shit in the morning."[29]

The real danger, however, is not just from the "kick in the air" from poop. The Conference on Environmental Health Impacts of Concentrated Animal Feeding Operations found that "Airborne contaminant emissions emanating from concentrated animal feeding operations (CAFOs) include toxic gases and particulates."[30] Nasty stuff that can kill you. Often, the workers on the farms are the ones most directly impacted by air emissions.[31] Why is this not an Occupational Safety and Health Administration (OSHA) violation?[32] Again, rural regulations are different for urban or other types of industrial operations. Studies from almost 40 years ago raised alarms about his crisis for workers.[33]

So why do the farm owners not do something? If things are so bad down on the farm, then certainly those that live there would make things better. Right? Well, the answer here is very simple. CAFOs are owned by absentee and are typically leased or contracted to poorer workers or managers.[34] The problem of soil management, air and water pollution, and malodors just do not impact them.[35] This is the classic "out-of-sight-out-of-mind" enigma. The owners do not live there. They live elsewhere.

However, beyond just the stink, compromised air quality from CAFOs extend into nearby communities, including schools, according to research.[36] How can little Billy or Sally learn or develop fully under such conditions? Again, quality of life on the farm and for communities nearby suffers due to CAFOs. The complexity of monitoring air emissions from CAFOs was part of an American Society of Agricultural and Biological Engineers study.[37]

The results of almost all peer-reviewed, independent research: CAFOs smell awful and pollute the water.

So, how did we get here?

Since 1965, CAFOs with open-pit lagoons have been considered the "technology" for waste treatment.[38] According to the formula, "A loading rate of 275 feeder hogs per acre at an average depth of 5 feet provides for odor-free operation with a minimum of maintenance and a long operational life. The total containment design without an overflow or discharge eliminates the necessity for a receiving stream or secondary treatment but may require a special make-up water source."[39] Some began to doubt this right away.[40] Adapting human waste (diluted with minimal nutrients) technologies for pig waste (concentrated with high nutrient content) was just part of the problem.[41] Simply stated, urban sewer plants could never manage the biological and chemical loading from a CAFO. Within a couple of years, sanitary engineers (today called wastewater engineers), said that the lagoon technology was not "state-of-the-art" and would become a "national problem."[42]

They were right. But there are more than just ecological fears.

Industrial pig farming not just stinks, again, it has health consequences for the nearby communities.[43] In the Pew Commission report (2008), Industrial Farm Animal Production, found that "public health concerns associated with CAFOs include heightened risks of pathogens passed from animals to humans; the emergence of microbes resistant to antibiotics and antimicrobials, due in large part to the widespread use of antimicrobials for nontherapeutic purposes, food-borne disease; worker health concerns; and dispersed impacts on adjacent community at-large."[44]

Further research demonstrates that exposure to ammonia and other pollutants (i.e., volatile organic compounds, particulate matter, nitrous oxide, hydrogen sulfide) is greater among poorer minorities than middle-class whites.[45] A North Carolina report (2013) linked CAFOs with calls for environmental justice.[46] These are more than just "nuisance" issues. Long-term mental and health impacts require attention for federal and state officials.[47] Making such improvements would require attention to larger issues such as the socio-economic factors of poverty and exposure of pollutants (not just from pigs).[48] In an outstanding health, poverty, and race analysis, researchers found "There are 18.9 times as many hog operations in the highest quintile of poverty as compared to the lowest; however, adjustment for population density reduces the excess to 7.2. Hog operations are approximately 5 times as common in the highest three quintiles of the percentage nonwhite population as compared to the lowest, adjusted for population density. The excess of hog operations is greatest in areas with both high poverty and high percentage nonwhites."[49]

Canadian researchers noted that the Pew report was accurate but should be increasingly applied beyond the fruited plains of Iowa and the tar-heeled hills of North Carolina. Increased demand for animal products has expanded pig production and therefore acerbated the public health and environmental woes associated with CAFOs.[50] But the US and Canada were not alone.

All over Europe, people were beginning to express concerns. The Netherlands has seen a remarkable shift towards IFAPs that is nothing short of an environmental and animal health "tragedy."[51] Due to the proximity of urban and rural communities in Europe, agricultural management (or mis-management) can radically alter air, water, and soil quality. Leaking lagoons and overflow into surface waters can quickly contaminate precious ground-water.[52] However, pollutants, such as ammonia, are not the only concerns. As has been the case in the US, Dutch pigs are loaded-up with antibiotics that researchers have found infect humans with more virulent diseases.[53] Italian researchers looked at total environmental costs from piglets to pigs to hogs.[54] Further research has targeted waste reduction and waste reuse tech-niques; with limited success and always with higher productions costs.[55]

Yet, the quality-of-life issue, as in the US, cannot be ignored. Unlike American farms that can be located far from populations (and their absen-tee owners), European CAFOs are only a few miles (kilometers) down the road. Odor control is a major concern. In fact, the European Union (EU) has a substantial body of regulations for odor pollution and impact.[56] The first EU regulations on odor were promulgated in the 1970s. In the Cambridge Environmental Research Consultants report (2015), various models for detec-tion, control, and diffusing of odors connected to public health and quality of life issues were detailed. The factors for assessing the impact of odors included: "frequency, intensity, duration, odor unpleasantness (or offensive-ness), and location."[57] While mercaptan (the added smell to natural gas so that people can detect a gas leak in their homes) is ranked at the top on this scale, odors from CAFOs are close, especially pig farms. Germany studies found that people miles away from CAFOs still were impacted by the odors and experienced diminished health.[58]

Alternatives to lagoons are plentiful. When farms were smaller, excess nutrients in pig waste could be land-applied or sprayed across fields. Crops could easily assimilate nutrient rich (nitrogen and phosphorus) waste. Today, with the ubiquitous IFAP models, nutrient uptake is just not possible.[59] It overloads the ecosystem. You can see the scorched fields where too much waste has been applied. Then, the rain washes excess nutrients into receiv-ing waters and kills fish and plant life. Europeans have looked at ways of managing the air emissions better due to the proximity of cities to farms. Here, researchers have analyzed air-scrubbers, like the ones on smokestacks at factories.[60] Technical limitations due to decentralized emissions is a con-straint to implementation. Where would you install such a system? There is no "smokestack or pipe" to scrub!

So, why do we have this terrible system? It is cheap. It keeps the price of pork artificially low. Others subsidize the pollution. The real cost of pork on the environment is what economists call an "externality" and is not part of direct costs to consumers.[61] European researchers began tracking these costs better so that a "true" price of pork could be determined.[62] To be fair, few industries account for all the externalities associated with production.

Yet, the focus here remains ecological. The EPA recognized this issue was beyond an air emissions problem. As early as 2002, the EPA released a shocking report that confirmed what many suspected: CAFOs were the major cause of water pollution in the US.[63] One interesting finding was that CAFOs generated six times the manure that humans did (which was regulated under the CWA).[64] Pollutants (including too many nutrients) in rivers, streams, and creeks deplete the water of oxygen disrupting aquatic systems. According to a Congressional report, "Excess nutrients cause fast-growing algae blooms that reduce the penetration of sunlight in the water column and reduce the amount of available oxygen in the water, thus reducing fish and shellfish habitat and affecting fish and invertebrates."[65]

The impact on water, as discussed above, remained in a major regulatory and bureaucratic quagmire. The federal conflict between the USDA and the EPA meant little actual regulation occurred. The problem was that the CWA regulated "point-source" discharges (think: end of a pipe), and CAFOs were considered "non-point source" discharges. Court decisions in the 1990s (as mentioned above) found that CAFOs were essentially large industries operating without permits.[66] Thus, the EPA began reviewing policies and regulations in an attempt to reign in the massive CAFOs that were compromising water quality in our nation's "navigable water."[67] So, in February 2003, the EPA revised the CWA to include CAFOs under the NPDES permitting regime. The expansion of the *Code of Federal Regulations* (CFR) now included the requirement for a discharge permit for CAFOs.[68] Would states enforce the CAFO requirement? Would all IFAPs get an NPDES permit?

For CAFOs, the law states, effluent (water discharge) limitations must use the "best practicable control technology currently available (BPT)" and "there must be no discharge of manure, litter, or process wastewater pollutants into waters of the U.S. from the production area."[69] In the EPA's 2004 Water Quality Inventory report (2009), evidence indicated that "excess algal growth alone is among the leading causes of impairment in lakes, ponds, and reservoirs, and that agricultural activities are among the top sources of lake impairments."[70]

Thus, the EPA (using the 1974 CWA standard of zero discharges for unpermitted facilities) created a three-tiered structure for CAFOs:

- The facility is a CAFO if it holds more than 1,000 animal units.
- If the facility holds from 300 to 999 animal units, the facility is a CAFO if pollutants are discharged from a manmade conveyance or are discharged directly into waters passing over, across, or through the site.

- Animal feeding operations that include fewer than 300 animal units may be designated as CAFOs if EPA or the permitting authority determines that the facility contributes significantly to water pollution.

The results have not been what many hoped. The CWA and the NPDES permitting system have made minimal difference as CAFOs continued to use open pit lagoons designed to leak or overflow.[71] Very few CAFOs actually sought a permit. There were exceptions for 24 rain events (thus the lagoons were allowed and expected to overflow), the NPDES permit did not demand a higher standard of technology; just the BPT. As a note, compare the CAA that has the highest threshold for technology regardless of cost. Groundwater concerns remained. Odor and nutrients were not managed either.[72] And for older CAFOs where solid waste has accumulated for years in the bottom of lagoons, the breakdown of the solids portion caused additional air emissions, and noxious gases are more than just an odor problem.[73]

So, the EPA attempted to step in. Yet, many claimed that this was another unfunded mandate placed on agricultural states by a "Big Brother" federal government. Who would pay the millions in management, compliance, and technology costs? Agricultural states are often not flush with cash. According to a Congressional report, "For state agencies that implement the NPDES permit program, the principal existing source of financial assistance is grants under Section 106 of the *Clean Water Act*, which states already use for various activities to develop and carry out water pollution control programs. States currently use Section 106 grants, supplemented by state resources, for standard setting, permitting, planning, enforcement, and related activities."[74]

There can be no doubt that farmers must be good stewards of the land to ensure a future for their way of life. No farm can destroy nature and expect to harvest a bountiful crop. Antagonistic ecological practices found with CAFOs foul the air, pollute the water, and taint the soil. They damage public health. But absentee owners often could care less. They do not live there. Reducing harmful air emissions and permitting water discharges should never have been the goal. It is but one step towards fixing a horrendous idea.

A dreadful idea that created an international environmental calamity. CAFOs had to find a solution. Big industry farming lobbyists had been able to shield IFAP for decades. Now, they were being hit from all sides. Neighbors everywhere were expressing concerns regarding odor and their quality of life.[75] Environmental activists were demanding cleaner air in Michigan.[76] Everyone seemed outraged about water pollution.[77] Environmental justice in North Carolina was another piece of the puzzle.[78] Animal rights activists wanted better, more sustainable, and humane conditions for pigs.[79]

That was the opportunity that presented itself.

Could there be an alternative "technological fix"? Certainly a $300 billion worldwide industry (today it is more than $400 billion) could afford to do what was ecologically friendly or sustainable?[80] Maybe. But at what cost to the industry and the consumer? US pig production is a huge international business. That business includes feed stores, veterinarians, fuel delivery, insurance and risk management, and dozens of others. The World Trade Organization (WTO) established an Agriculture Committee and considers import-export relations to be linked to transnational agreements and food security.[81]

Bilateral trade for the US pig industry is noteworthy. The biggest states for pork production are Iowa and North Carolina with Minnesota, Texas, and Oklahoma distantly on the radar. Could other states look for opportunities to enter the global pig market?[82] Pork consumption ranks behind beef and poultry in the US, but many countries, like Asian states, consume much more.[83] In fact, China, Mexico, Japan, and Canada imports more than 75 percent of US pork.[84] China alone represents more than 50 percent of all consumption worldwide and they export almost none.[85] Mexico is critical because they provide piglets to farms all over America.

Yet, CAFOs are not sustainable farming. The current system of ecological stewardship for CAFOs does not fully account for all costs of waste management and pushes responsibility for clean-up to another generation. According to researchers, "Satisfying the demand for food is already driving changes in crop and livestock production methods that may have profound environmental effects."[86] When you contemplate about the difficulties of climate change adaptation and mitigation policies and techniques, CAFOs come up very short in meeting those demands.[87]

Yet, in the late 1990s, climate change compliance matters were hardly the driving force behind any proposed regulatory modifications. There was a multi-billion-dollar industry being limited because of poor waste management. Industry growth was being hemmed in by environmental limits. Odor may have driven complaints but the issue was much deeper, and more complex.

Why? What was the root cause of the problem? Were there just too many pigs? Or, was the problem merely one of too much waste in too little space? What alternatives were there? Maybe there was a technological "fix"? Could there be a way to capture the energy value of the waste and make biofuel? Or, recycle or compost the solids part of the waste? Maybe the liquid part of the waste could be land-applied? If the odor problem was "solved" could that allow CAFOs to just continue on "business-as-usual"?

Pigs and me were about to run headlong into one another.

I was about to complete my PhD with a focus on environmental law and policy. I had recently accepted a research position with an EPA-funded think tank at the University of New Orleans. I had an amazing set of colleagues from a diversity of backgrounds: engineering, chemistry, and biology. I was teaching across town at Tulane University. Despite having very little expertise in animal waste management, I wanted to make a difference.

This project could result in changed regulations, policy, technology, and improve the air and water quality for millions of people. It also seemed very lucrative to whomever solved this.

I did not need to be "sold" on the idea.

International travel . . . high-level government officials . . . funded-research . . . publications . . . a patent . . . tenure!

I was hooked from the very beginning.

I had "played around" with environmental projects in South America and Europe. I had written my dissertation on research in Poland and the Czech Republic. But this was an opportunity to do more than observe or provide analysis to policymakers. This time, I would be at the table making decisions (and money).

In my "big yellow taxi," I would save the environment.[88]

I was staunchly dedicated to a righteous cause that would transform the world!

For the next decade, I lived in a world of pig waste; surrounded by an assortment of characters that most people have never heard of until now. There was a religious-like zeal among this group. We had our leader, a cast of supporting technicians, lawyers, connected international businesspeople with the ability to make telephone calls and change public-policy.

About a dozen of us sacrificing everything. A sole-purpose and a fervent commitment to passionately pursue pig poop solutions.

We all knew, as did anyone analyzing this problem, managing pig waste from CAFOs in lagoons was a terrible idea. There was decades of studies, university, and corporate documents supporting the need to do something different.

We assumed that having a great, novel idea would propel us to ecological glory.

Our ignorance and gullibility far eclipsed anything a CAFO could produce.

I do not believe we facilitated the industrial lie of pig lagoons.

But we were still unknowing abettors to hog wash.

NOTES

1. Ruhl, J. B. (2000). Farms, their environmental harms, and environmental law. *Ecology LQ, 27*, 263.

2. Phillips, S. T. (2007). *This land, this nation: Conservation, rural America, and the New Deal*. Cambridge University Press.

3. Bonnie, R., Diamond, E. P., & Rowe, E. (2020). *Understanding rural attitudes toward the environment and conservation in America*. Nicholas Institute for Environmental Policy Solutions, Duke University.

4. Trusts, P. C., & Hopkins, J. (2008). *Putting meat on the table: Industrial farm animal production in America*. A Report of the Pew commission on industrial Farm Animal Production.

5. Peters, K. A. (2010). Creating a sustainable urban agriculture revolution. *Journal of Environmental Law and Litigation, 25*, 203.

6. US Government Accountability Office. Concentrated animal feeding operations: EPA needs more information and a clearly defined strategy to protect air and water quality from pollutants of concern. Located at https://www.gao.gov/products/gao-08-944. Accessed 6 June 2022.

7. Spellman, F. R., & Whiting, N. E. (2007). *Environmental management of concentrated animal feeding operations (CAFOs)*. CRC Press.

8. Chastain, J. P., Camberato, J. J., Albrecht, J. E., & Adams, J. (1999). *Swine manure production and nutrient content. South Carolina confined animal manure managers certification program*. Clemson University, SC, 1–17.

9. Mallin, M. A., McIver, M. R., Robuck, A. R., & Dickens, A. K. (2015). Industrial swine and poultry production causes chronic nutrient and fecal microbial stream pollution. *Water, Air, & Soil Pollution, 226*(12), 1–13.

10. Barker, J. C., & Zublena, J. P. (1995). *Livestock manure nutrient assessment in North Carolina. Final Report*. North Carolina Agricultural Extension Service, North Carolina State University.

11. Bradford, S. A., Segal, E., Zheng, W., Wang, Q., & Hutchins, S. R. (2008). Reuse of concentrated animal feeding operation wastewater on agricultural lands. *Journal of Environmental Quality, 37*(S5), S-97.

12. Harden, S. L. (2015). Surface-water quality in agricultural watersheds of the North Carolina Coastal Plain associated with concentrated animal feeding operations (No. 2015–5080). US Geological Survey.

13. Hribar, C. (2010). Understanding concentrated animal feeding operations and their impact on communities.

14. Mitloehner, F. M., & Calvo, M. S. (2008). Worker health and safety in concentrated animal feeding operations. *Journal of Agricultural Safety and Health, 14*(2), 163–187.

15. Liu, X., Zhang, W., Hu, Y., Hu, E., Xie, X., Wang, L., & Cheng, H. (2015). Arsenic pollution of agricultural soils by concentrated animal feeding operations (CAFOs). *Chemosphere, 119*, 273–281.

16. Goldberg, A. M. (2016). Farm animal welfare and human health. *Current Environmental Health Reports*, 3(3), 313–321. Some states have asked the ethical questions regarding such practices. Doonan, G., Appelt, M., & Inch, C. (2009). Role of legislation in support of animal welfare. *The Canadian Veterinary Journal, 50*(3), 233.

17. Hobbs, P. J., Misselbrook, T. H., & Cumby, T. R. (1999). Production and emission of odours and gases from ageing pig waste. *Journal of Agricultural Engineering Research, 72*(3), 291–298.

18. Burkholder, J., Libra, B., Weyer, P., Heathcote, S., Kolpin, D., Thorne, P. S., & Wichman, M. (2007). Impacts of waste from concentrated animal feeding operations on water quality. *Environmental Health Perspectives, 115*(2), 308–312.

19. Ben, W., Pan, X., & Qiang, Z. (2013). Occurrence and partition of antibiotics in the liquid and solid phases of swine wastewater from concentrated animal feeding operations in Shandong Province, China. *Environmental Science: Processes & Impacts, 15*(4), 870–875.

20. Ndegwa, P. M., Hristov, A. N., Arogo, J., & Sheffield, R. E. (2008). A review of ammonia emission mitigation techniques for concentrated animal feeding operations. *Biosystems Engineering, 100*(4), 453–469.

21. Rowan, J. P., Durrance, K. L., Combs, G. E., & Fisher, L. Z. (1997). The digestive tract of the pig. Animal Science Department, Florida Cooperative Extension Service, Institute of Food and Agricultural Sciences, University of Florida, Gainesville, Document AS23, 1(4), 1–7.

22. Manners, M. J. (1976). The development of digestive function in the pig. *Proceedings of the Nutrition Society, 35*(1), 49–55.

23. Sousan, S., Iverson, G., Humphrey, C., Lewis, A., Streuber, D., & Richardson, L. (2021). High-frequency assessment of air and water quality at a concentration animal feeding operation during wastewater application to spray fields. *Environmental Pollution, 288*, 117801.

24. Rosov, K. A., Mallin, M. A., & Cahoon, L. B. (2020). Waste nutrients from US animal feeding operations: Regulations are inconsistent across states and inadequately assess nutrient export risk. *Journal of Environmental Management, 269*, 110738.

25. In response to Jules saying, "Pigs are filthy animals. I don't eat filthy animals." Vincent provides his memorable line. This is from the movie *Pulp Fiction* (1994).

26. Natural Resources Defense Council v. Reilly, U.S. District Court, D.C., Civ. Action No. 89–2980, April 23, 1991.

27. National Research Council. (2003). Air emissions from animal feeding operations: Current knowledge, future needs.

28. Kirkhorn, S. R. (2002). Community and environmental health effects of concentrated animal feeding operations. *Minnesota Medicine, 85*(10), 38–43.

29. Apologies to Lieutenant Colonel Bill Kilgore (Apocalypse Now).

30. Bunton, B., O'Shaughnessy, P., Fitzsimmons, S., Gering, J., Hoff, S., Lyngbye, M., . . . & Werner, M. (2007). Monitoring and modeling of emissions from concentrated animal feeding operations: overview of methods. *Environmental Health Perspectives, 115*(2), 303–307.

31. Mitloehner, F. M., & Calvo, M. S. (2008). Worker health and safety in concentrated animal feeding operations. *Journal of Agricultural Safety and Health, 14*(2), 163–187.

32. Kelly-Reif, K., & Wing, S. (2016). Urban-rural exploitation: An underappreciated dimension of environmental injustice. *Journal of Rural Studies, 47*, 350–358.

33. Bongers, P., Houthuijs, D., Remijn, B., Brouwer, R., & Biersteker, K. (1987). Lung function and respiratory symptoms in pig farmers. *Occupational and Environmental Medicine, 44*(12), 819–823.

34. Petrzelka, P., Ma, Z., & Malin, S. (2013). The elephant in the room: absentee landowner issues in conservation and land management. *Land Use Policy*, 30(1), 157–166.

35. Ulrich-Schad, J. D., Babin, N., Ma, Z., & Prokopy, L. S. (2016). Out-of-state, out of mind? Non-operating farmland owners and conservation decision making. *Land Use Policy, 54*, 602–613.

36. Barrett, J. R. (2006). Hogging the air: CAFO emissions reach into schools. A wonderful study linked CAFO emissions to higher blood pressure levels. Wing, S., Horton, R. A., & Rose, K. M. (2013). Air pollution from industrial swine operations and blood pressure of neighboring residents. *Environmental Health Perspectives, 121*(1), 92–96.

37. Parker, D. B., Caraway, E. A., Rhoades, M. B., Cole, N. A., Todd, R. W., & Casey, K. D. (2010). Effect of wind tunnel air velocity on VOC flux from standard solutions and CAFO manure/wastewater. *Transactions of the ASABE, 53*(3), 831–845.

38. Clark, C. E. (1965). Hog waste disposal by lagooning. *Journal of the Sanitary Engineering Division, 91*(6), 27–42.

39. Ibid.

40. Hart, S. A., & Turner, M. E. (1965). Lagoons for livestock manure. *Journal (Water Pollution Control Federation)*, 1578–1596.

41. Neel, J. K., & Hopkins, G. J. (1956). Experimental lagooning of raw sewage. *Sewage and Industrial Wastes*, 28(11), 1326–1356.

42. Loehr, R. C. (1969). Animal wastes—a national problem. *Journal of the Sanitary Engineering Division, 95*(2), 189–222.

43. Greger, M., & Koneswaran, G. (2010). The public health impacts of concentrated animal feeding operations on local communities. *Family and Community Health*, 11–20.

44. Pew Commission on Industrial Farm Animal Production. Located at https://www.pcifapia.org/. Accessed 2 June 2022.

45. Mirabelli, M. C., Wing, S., Marshall, S. W., & Wilcosky, T. C. (2006). Race, poverty, and potential exposure of middle-school students to air emissions from confined swine feeding operations. *Environmental Health Perspectives, 114*(4), 591–596.

46. Nicole, W. (2013). CAFOs and environmental justice: The case of North Carolina.

47. Heederik, D., Sigsgaard, T., Thorne, P. S., Kline, J. N., Avery, R., Bønløkke, J. H., . . . & Merchant, J. A. (2007). Health effects of airborne exposures from concentrated animal feeding operations. *Environmental Health Perspectives, 115*(2), 298–302.

48. Donham, K. J., Wing, S., Osterberg, D., Flora, J. L., Hodne, C., Thu, K. M., & Thorne, P. S. (2007). Community health and socioeconomic issues surrounding concentrated animal feeding operations. *Environmental Health Perspectives, 115*(2), 317–320.

49. Taylor, D. A. (2001). From pigsties to hog heaven? *Environmental Health Perspectives, 109*(7), A328–A331.

50. Gyles, C. (2010). Industrial farm animal production. *The Canadian Veterinary Journal, 51*(2), 125.

51. Ashwood, L. (2012). Daniel Imhoff (Ed): The CAFO reader: the tragedy of industrial animal factories. *Agriculture and Human Values, 29*(3), 427–428.

52. Szekeres, E., Chiriac, C. M., Baricz, A., Szőke-Nagy, T., Lung, I., Soran, M. L., . . . & Coman, C. (2018). Investigating antibiotics, antibiotic resistance genes, and microbial contaminants in groundwater in relation to the proximity of urban areas. *Environmental Pollution, 236*, 734–744.

53. West, B. M., Liggit, P., Clemans, D. L., & Francoeur, S. N. (2011). Antibiotic resistance, gene transfer, and water quality patterns observed in waterways near CAFO farms and wastewater treatment facilities. *Water, Air, & Soil Pollution, 217*(1), 473–489.

54. Pirlo, G., Carè, S., Della Casa, G., Marchetti, R., Ponzoni, G., Faeti, V., . . . & Falconi, F. (2016). Environmental impact of heavy pig production in a sample of Italian farms. A cradle to farm-gate analysis. *Science of the Total Environment, 565*, 576–585.

55. Pinotti, L., Luciano, A., Ottoboni, M., Manoni, M., Ferrari, L., Marchis, D., & Tretola, M. (2021). Recycling food leftovers in feed as opportunity to increase the sustainability of livestock production. *Journal of Cleaner Production, 294*, 126290.

56. European Union. Research & Innovation Report. Analysis of existing regulation in odour pollution, odour impact criteria 1. September 2019.

57. Price, C. (2015). *Odour regulations in Europe—different approaches*. CERC Publication.

58. Osterberg, D., & Merchant, J. (2017). CAFOs and the Diminished Defence of Public Health.

59. McLaughlin, M. R., Fairbrother, T. E., & Rowe, D. E. (2004). Nutrient uptake by warm-season perennial grasses in a swine effluent spray field. *Agronomy Journal, 96*(2), 484–493.

60. Costantini, M., Bacenetti, J., Coppola, G., Orsi, L., Ganzaroli, A., & Guarino, M. (2020). Improvement of human health and environmental costs in the European Union by air scrubbers in intensive pig farming. *Journal of Cleaner Production, 275*, 124007.

61. Nguyen, T. L. T., Hermansen, J. E., & Mogensen, L. (2012). Environmental costs of meat production: the case of typical EU pork production. *Journal of Cleaner Production, 28*, 168–176.

62. Dorca-Preda, T., Mogensen, L., Kristensen, T., & Knudsen, M. T. (2021). Environmental impact of Danish pork at slaughterhouse gate–a life cycle assessment following biological and technological changes over a 10-year period. *Livestock Science, 251*, 104622.

63. Environmental Protection Agency. National Service Center for Environmental Publications. Environmental and Economic Benefit Analysis of Final Revisions to the National Pollutant Discharge Elimination System Regulation and the Effluent Guidelines for Concentrated Animal Feeding Operations, December 2002.

64. Ibid.

65. Congressional Research Service. Animal Waste and Water Quality: EPA Regulation of Concentrated Animal Feeding Operations (CAFOs). 16 February 2010. Claudia Copeland, Specialist in Resources and Environmental Policy.

66. Martin Jr, J. H. (1997). The clean water act and animal agriculture. *Journal of Environmental Quality, 26*(5), 1198–1203.

67. Parry, R. (1998). Agricultural phosphorus and water quality: A US Environmental Protection Agency perspective. *Journal of Environmental Quality, 27*(2), 258–261.

68. CFR. Title 40, Chapter 1, Subchapter D, part 122. Appendix A.

69. CFR. § 412.31, (a) (1).

70. EPA. National water quality inventory: Report to Congress for the 2004 reporting cycle, January 2009, EPA-841-R-08–001, pp. 18–19.

71. Laitos, J. G., & Ruckriegle, H. (2012). The clean water act and the challenge of agricultural pollution. *Vermont Law Review*, 37, 1033.

72. US General Accounting Office, Livestock agriculture: Increased EPA oversight will improve environmental program for concentrated animal feeding operations, January 2003, GAO-03–285, p. 7. An interesting note here is that the EPA regulated animals differently based on their weight. Chickens, cows, and pigs had different standards.

73. Hobbs et al.

74. CRS. The costs for state agency management was not worked out here. States would have "to carry out their responsibilities under the revised rules without reducing resources for other important activities."

75. Doane, M. (2014). Politics and the family farm: When the neighbors poison the well. *Anthropology Now, 6*(3), 45–52.

76. Zande, K. (2008). Raising a Stink: Why Michigan CAFO Regulations Fail to Protect the State's Air and Great Lakes and Are in Need or Revision. *Buffalo Environmental Law Journal, 16*, 1.

77. Sato, A. (2005). Public Participation and *Access to Clean Water: An Analysis of the CAFO Rule. Sustainable Development and Law and Policy, 5*, 40.

78. Ball-Blakely, C. (2018). CAFOs: Plaguing North Carolina communities of color. *Sustainable Development Law & Policy, 18*(1), 3.

79. Walton, L., & Jaiven, K. K. (2020). Regulating CAFOs for the Well-Being of Farm Animals, Consumers, and the Environment. *Environmental Law Report, 50*, 10485.

80. Transparency Market Research. Hog Production and Pork Market. Located at https://www.transparencymarketresearch.com/hog-production-pork-market.html #:~:text=The%20global%20hog%20production%20and,a%20CAGR%20of%20~2 %25. Accessed 20 May 2022. According to CRS report (2015), as cited above, "The proposed and final rules also presented EPA's estimates of the costs of revised regulation. EPA estimated that the total incremental compliance costs for CAFOs is $326 million annually (pre-tax, 2001 dollars), consisting of $283 million for large CAFOs, $39 million for medium CAFOs, and $4 million for facilities that are designated as CAFOs. Federal and state permitting authorities were projected to incur $9 million per year in costs to implement the rules. Estimated annual

incremental costs of the proposed rules were $831-$930 million for CAFO operators, plus $6–8 million for permitting authorities (1999 dollars)." This is but a scratch on the $400 billion industry.

81. World Trade Organization. Agriculture. Located at https://www.wto.org/english/tratop_e/agric_e/agric_e.htm. Accessed 2 June 2022.

82. Orr, D. E., & Shen, Y. (2006). World pig production, opportunity or threat. In Indiana Farm Bureau's Midwest Swine Nutrition Conference, Indiana.

83. Davis, C. G., & Lin, B. H. (2005). Factors affecting US pork consumption. US Department of Agriculture, Economic Research Service.

84. USDA. Animal Products. Located at https://www.ers.usda.gov/topics/animal-products/hogs-pork/sector-at-a-glance/. Accessed 2 May 2022.

85. Schnitkey, G. (2013). Chinese and US pork consumption and production. *Farmdoc Daily*, 3.

86. Aneja, V. P., Schlesinger, W. H., Nyogi, D., Jennings, G., Gilliam, W., Knighton, R. E., . . . & Krishnan, S. (2006). Emerging national research needs for agricultural air quality. *Eos, Transactions American Geophysical Union, 87*(3), 25–29.

87. Tomas, K. A. (2018). Manure Management for Climate Change Mitigation: Regulating CAFO Greenhouse Gas Emissions Under the Clean Air Act. *U. Miami L. Rev., 73*, 531.

88. Apologies to Joni Mitchell.

Chapter 2

The Foul Wind Blows

It was September 11, 1999. Hurricane Floyd blasted the North Carolina coast like a bad dream and left behind damage like a slow divorce. The National Weather Service (NWS) reported top winds speeds along the coast at 138 miles per hour (mph) and inland gusts greater than 70 mph.[1] The storm surge carried a wall of water more than 15 feet high combined with tornadoes across communities ill-equipped for such wreckage. The pig farms in the North Carolina region would suffer hard.

Coastal communities prepare for wind and rain. But, as reported by the NWS, "The effects of Hurricane Floyd were exacerbated by Hurricane Dennis. Dennis slammed the North Carolina coast less than a month before Hurricane Floyd with over 10 inches of rain in areas of eastern North Carolina. The resulting rainfall saturated much of the soil in North Carolina and portions of Virginia. This helped to increase the flooding potential of Hurricane Floyd."[2] The foul wind blew more than an ugly storm on land. It stripped bare any pretense that the drainage pits (lagoons) employed by CAFOs could withstand such torrential rain.

Moore and Barnes (2004) detailed the "epic flood that ranks as the most widespread, destructive, and deadly natural disaster in North Carolina's history."[3] In their recollections, they provide context for a storm that no one had ever seen before in that part of the country. How could anyone imagine such destruction? They noted, "Sixty-six counties were declared disaster areas, damage estimates exceeded $6 billion, and there were fifty-two reported fatalities . . . More than sixty thousand homes were flooded . . . "[4]

Another reported added, Hurricane Floyd left "1.5 million people without power, 48,000 people in shelters, and hundreds of thousands of livestock dead. Many of the animals had no place to run for safety as they were confined inside of death cages already. In addition to the CAFO and home flooding, the storm submerged twenty-four wastewater treatment plants, destroyed seven dams, and caused over $150 million in damage to state highways and $75 million to bridges and drainage systems."[5] Sewer plants require a steady

17

food source of sludge for the bacteria to do their job. Excess water flushed the system, and the bacteria gets washed out to sea. Thus, no treatment occurs, and raw sewage slushes through streets on way to rivers, streams, and lakes.

The human damage cannot be overemphasized. The loss of personal property, crucial infrastructure, and human life was enormous.

Yet, again, the foul wind of Floyd left behind more than scattered debris and flooded communities. Floyd and his vile brother Hurricane Dennis unveiled the terrible effects of poorly conceived and ill-designed CAFO lagoons. As the rain fell and the waters washed through communities, lagoon levies broke and emptied their contents in neighboring rivers. Pig waste and sewage waters formed a toxic brew. A contaminated cocktail of pollution of the most lethal form.

Researchers found that "Hurricane Floyd impacted the Neuse River basin by inducing flooding that damaged and disabled hog lagoons and municipal wastewater treatment plants. Between approximately 53 and 325 million liters (14 and 86 million gallons) of untreated hog waste and between approximately 5.7 and 34.4 billion liters (1.5 and 9.1 billion gallons) of untreated municipal wastewater are projected to have entered the Neuse River basin, increasing the concentrations of nitrogen (N), phosphorus (P), and total solids."[6] Most people seeing the news would be unaware that the real ecological nightmare is not the damaged homes or displaced people. The sewage and slurry soup were feeding the most nasty and unsanitary elixir conceivable.

The pollution would render thousands of homes uninhabitable and created a potential health crisis due to exposure to this horrid waste concoction.[7] At least that was the expectation. Perhaps it was due to limited health care insurance among the poorest impacted, but one study found relatively few claims specifically associated with exposure to pig waste.[8] This was anticipated because of the extreme poverty along Floyd's path.

Still, something had to be done. The breach of pig lagoons was offensive and intolerable to anyone close to the coast. Many began to ask whether the state would now finally do something to change policies and support better "technology" than a leaky, smelly lagoon.[9] Politics aside, North Carolina was paying an enormous price to subsidize the pig industry. There had been many that had questioned lagoons for CAFOs before Floyd and Dennis trampled through the coast.[10] While industry farmers would assert that the costs would drive pork prices too high (and subsequently them out of business), Congress had provided a financial remedy three years earlier. States had a remedy in-place if they chose to utilize it.

The *Federal Improvement and Reform Act* (FAIR) authorized more than $1.3 billion for environmental technologies specific to CAFOs.[11] It is important to make a distinction once again. While the *Clean Air Act* (CAA) is based on the best technology regardless of cost (protecting human health toxic air is critical), the *Clean Water Act* (CWA) requires cost-benefit analysis (and that means jobs) prior to promulgation. Would FAIR be driven by health protection or cost-benefit analysis?

Thus, as researchers found, FAIR never constituted a national standard for all technology. Instead, it allowed for a minimum set of requirements that was hardly better than the present system.[12] FAIR never lived up to its name. Especially for minorities in heavy CAFO communities. So-called agricultural experts offered little here as spraying lagoon water on crops still represented the potential for run-off following even the slightest downpour.[13] Neither the CAA nor the CWA would have relied upon such a "technology" as a basis for improvement. It seemed that state legislatures and the CAFO industry had outmaneuvered Congress.

The problem, though, went beyond the hazardous waters (and the disgusting air). The CAFO system meant pigs were packed into small cages with limited mobility. As the winds howled and the rain dropped, the helpless pigs could not escape.[14] According to initial reports, more than 100,000 dead pigs floated like pork flotsam and jetty throughout the floodwaters of North Carolina.[15] But this could have been prevented. Even just a more humane method of animal farming could have allowed the pigs to swim to safety. As noted, "Most alarming of all is that this had all been predicted and anticipated as an outcome of industrial pig production that was virtually unchecked by legislation while it was already destroying the economic and social lives of small producers on family farms."[16]

The images of dead pigs rotting in cities and towns was shocking. Some reported seeing pigs rotting on top of barns.[17] I remember seeing the destruction of Floyd. Having lived on the Louisiana Gulf Coast, I was aware of what a hurricane could leave behind. I was alarmed to see dead pigs floating through the floodwater currents downstream to infect and pollute other communities. "Dead boxes" were placed alongside the roads to dump carcasses.[18] It was a real horror show. The rancid, rotting flesh of these hogs reeked a malodourous stench that floated across valleys like some horrid phantasm.

This was no delusion, though.

Many were looking for someone to blame.

So, it was an opportunity to take (legitimate) shots at industry giants like Smithfield Foods.[19] Smithfield Foods was the world's largest, and most profitable pig producer. They were founded shortly after World War Two in Virginia and have a multi-billion-dollar footprint spanning the globe.[20] Probably anyone that has ever eaten BBQ along America's East Coast has tasted a savory Smithfield hog. The argument over mustard or tomato sauce or even a dry rub brings out the swine squabble even among quiet folks. The numbers for Smithfield are staggering. Annually they butcher more than

25 million pigs and produce hundreds of thousands of tons of waste.[21] But other producers, especially those impacted by Floyd were called out. North Carolina pig farms like Carol and Murphy, which Smithfield Foods later bought out, would face criticism. This time the condemnation would enlarge and combine race and environmental outrage.[22]

Is having scrumptious BBQ worth the costs?

Should hog farming be more sustainable?

The industry was in trouble.

Really?

Not so fast!

Would politicians allow a consistent source of funds to dry-up? The Institute for Southern Studies calculated the political contributions of industrial pig operations and found something interesting. According to their report, "Although hogs are raised primarily in eastern North Carolina, more than half the current legislators (92 of 170) got money from the pork industry's executives, lobbyists, political action committees, hog farmers, and their families. Hog operators rank among the state's biggest individuals contributors. Two of the four families that gave the most money to members of the 1995 General Assembly are headed by Wendell H. Murphy and William H. Prestage, two of the state's (and the nation's) largest hog farm operators."[23]

That was in 1995.

What about more recent contributions?

In 2017, "Big Pork . . . contributed more than $272,000 to the campaigns" mainly to support candidates favoring protecting the industry.[24] Is that so surprising, though? Businesses make donations and get access to politicians so they can get favorable rules and regulations and make higher profits. The system is set up that way, right? Should it be? How can poor people with limited resources ever effect positive change?

One such bill[25] had a retroactive provision so that any past lawsuits would become quashed. Conservation, religious, and community groups cried foul and, once again, linked this to environmental justice and racism. The bill was called *Agriculture and Forestry Nuisance Remedies*. According to one report, "North Carolina's industrial hog farms are notoriously toxic, storing millions of gallons of feces and urine in open-air cesspools. When those pools fill up, the hog waste sprays into the air, activists say, and can even make its way into peoples' homes. The stenches and fumes can be so repulsive that people living near the farms will quarantine themselves inside or only step outside briefly. In short, nearby residents say, it's a horrible and nauseating way to live."[26]

One resident claimed, "I live in the middle of around twenty hog farms. And they smell so bad you can hardly come outside most of the time. And we just want to try to stop the pollution. The hog farms are polluting our areas and we want to try to stop them."[27] Is that really an unreasonable request?

Another resident and member of a local Waterkeeper organization,[28] added, "This is about race. This bill has got to go. It's not about protecting the people; it's about protecting the industry."[29] The national attention from Hurricane Floyd meant environmental organizations had fresh ammunition to hurl at CAFOs.

Pig waste causing environmental damage and impacting poor people of color? So why does nothing seem to get fixed here? Some would argue that pork profits are just more important than poor people. Big money can deflect and dissuade politicians from doing their jobs. Where do poor people of color have a voice that shouts louder than money?

This all sounds too familiar.

But agricultural lobbying has been going on for decades. In states with large industrial farming, what would you expect? The issue is not so much whether industry of any kind makes campaign donations. The issue is buying influence.[30] Every imaginable sector contributes to campaigns. Does this allow businesses to secure votes? For merely access?[31] Allegations, however, are not proof. Inuendo is not evidence. Specifically connecting money and votes has been the subject of many studies for years.[32]

Correlation is not causation.

It just seems odd.

The reeking stench of political mendacity may have eclipsed the foul odors of dead pigs. And it would continue to linger well after the last pig was pulled from the rubble. Eventually those pigs decomposed, the lasting effect of political corruption damaged institutions.

Hurricane Floyd brought the pig industry into the living rooms of Americans. So, what changed? Is the industry more sustainable? Eco-friendly? Less inhumane? What do you think?

In a Coastal Review article, it was claimed by Rick Dove, the retired Neuse Riverkeeper who is now an associate with the Waterkeeper Alliance, "You would think as a result of Floyd that major changes would have been made. I regret to inform you, they have not."[33] One writer offers his lamentations in a poem,

> *Near metal sheds, a hog farm's effluent*
> *Fans Pepto-Bismol pink in the current*
> *Unfurling its nasty plume downstream.*
> *Pigs rosy as in a dream*[34]

It would take more than poetry to change a multi-billion-dollar industry. Lots of hard work, brains, and massive connections would be needed to stem the slurry tide. Unless business and politics came together no solution, regardless of the technology or cost-benefit was going to be adopted. The "solution" would need to check all the boxes: technology, costs, law, and the environment.

By the time this issue hit my radar, a couple of years before Hurricane Floyd, I was already working at the University of New Orleans Urban Waste Management & Research Center (UWMRC), an EPA funded think tank and policy group.[35] The UWMRC would write, submit, and receive grant funding, mainly from the EPA, and other federal, state, and local government agencies (and sometimes private companies or non-profits). This was a collection of bright, talented professors, researchers, and graduate students from environmental engineering, biology, and chemistry backgrounds.

I was none of these things.

I was completing my PhD in political science. However, I had taken some graduate environmental engineering classes and many other environmental classes. My experience in working with the UWMRC allowed me to produce and direct some environmental documentary films on solid waste and conduct research alongside these capable people. I was towards the end of another film series on water quality shot throughout the Gulf States (Florida to Texas) when I saw these dead pigs

My UWMRC colleagues were Dr. Al Knecht and Juan C. Josse. Dr. Knecht was an elderly scholar, researcher with an understanding across regulatory, technical, and legal sectors. He had served in various capacities for industry and government as a consultant. More than anything else, he was a professional and a mentor. He was kind and supportive. He was married for more than 64 years to Gretchen, and I never heard anyone say a harsh word against him.[36] His collaborative research spanned decades and his impact encompassed environmental sectors that brought various disciplines together to solve problems. He had the ability to analyze an environmental issue from every plausible perspective. Dr. Al Knecht died in 2019. I was very sad when I heard that he passed away. I still had so much to learn from him.

Juan Josse was Chilean from Ecuador. Josse stood over six feet tall, was blonde, blue-eyed, and often tan. He played the guitar, rode the waves on a surfboard, camped on the beach with his dog, and was a genius. The fact that he was a committed Socialist and often anti-American meant that he and I would frequently butt heads. He was impossible not to like. A wide smile full of white teeth and a brightness to his face showed an inner light. Maybe in another life he was a shaman or priest. He was the most ethical, moral person in which I have ever worked. Because of his looks, brains, and various skills, Josse was loved by everyone. Since I had lost my oldest and youngest brothers by this point, I looked to Juan as a brother. The loss of his friendship during this pig fiasco was unfortunate and disturbs me to this day. It is one of the most consequential losses of my life.

Juan's influence on me went beyond environmental matters. A few weeks earlier, I was asked to participate in an Indonesian project where native people were rioting and killing corporate and government officials. They were outraged because the mining was killing the Earth. His solution was plain, "John, the company should not be there. They are ruining these people's lives over some minerals dug from the Earth that belongs to everyone. No one has that right."

We were traveling across Lake Ponchartrain (New Orleans) on the world's longest bridge over water[37] to consult on a project for a growing parish that needed assistance with wastewater management. I had plenty of time to think about his comments. We listened to music and sang as we drove. As a note, whenever he said my first name, I knew that Juan was serious. I generally listened.

He was right. I could not participate in a "study" of indigenous people so that a corporation could use my data and analysis to further exploit them. That is not why I was developing my environmental assessment skills. At some point, I believe that everyone makes a choice for what they believe. It was my time to make such a decision.

Since we had arrived early, I asked if I could stop by my professor's house for a minute. The professor was Dr. Robert S. Jordan. He was the chair of my dissertation committee and a strong influence on me at the time. Professor Jordan worked at the United Nations and was known throughout Europe and Africa being an expert on the international civil service system and international organizations (IOs). He had pitched a project to a private company, and I simply said yes without thinking too much about the issues. We were going to provide research and assistance to this company facing the outrage of indigenous people. To me, it would be a great experience. I would make good money and gain more international knowledge.

Josse had thrown a principled crimp in my plans. I had been blinded by my desires to advance my career. I had never considered an opposing view. The ethical view. The moral view.

"Of course, it is on the way," he generously said.

"Thanks."

We stopped at Dr. Jordan's house. A two-story home in the suburbs of New Orleans. An unpretentious, but elegant dwelling. Simple tastes. Not garish. The kind of house with a formal living room that is never used and a livable kitchen where all meeting s were held. He had converted his garage into a private office. I knocked, Dr. Jordan greeted me, and entered while Juan stayed in the car. After a few pleasantries, I resigned from the project.

My professor was in shock.

"Why," he asked.

"I simply cannot work on a project with these people profiting from hurting so many disenfranchised people," I explained.

He just looked at me.

I apologized for letting him down and politely excused myself.

I felt that doing the ethical thing should never be postponed. A delay would be the same as acquiescing. I went back to the car where Juan was waiting. He had been fumbling with the radio.

"That did not take long," Juan said.

He read the expression on my face.

"I told Professor Jordan that I would not work with him in Indonesia on this project," I clarified.

"Really?"

"Yes. Now let's go to our meeting and get to work."

My relationship with Dr. Jordan never recovered. He chaired my dissertation committee and I needed his full support to make it through the snake-pit of university politics. I never got it. We spoke less and less after that day. He died more than 25 years later at the age of 92.[38] I knew that he was disappointed in me and could never understand. He was also a man of principles. We simply saw things differently. Among people with strong personalities and beliefs that can happen. I am certain that it caused a strain with other graduate students and me that lasts to this day.

So, Al, Juan, and I would look for ways to collaborate. It did not take long.

Jefferson Parish lies adjacent to New Orleans. There, someone else had been looking at CAFOs, and then later had seen the effects of Hurricane Floyd and all those dead pigs. Jordan Hill[39] was long and thin. He sat day after day watching the news and scheming up ways to make a fortune. His throne was a leather Lazy-Boy chair that when Hill reclined, he seemed to grow another foot in length. I rarely saw him wear shoes. Hill always had thick socks. He smoked cigarettes and drank black coffee. His breath was like a train engine from the 1800s. Occasionally, when he got up, he cooked and then retreated to the comfort of his chair. His living room was scattered with printouts and various newspaper clippings. He spoke in an older New Orleans accent and preferred slang to English, was a complete racist, homophobe, and a religious bigot. He seemed to be educated on trash TV, religious shows, and loved anything salacious.

Hill reached out to Dr. Knecht.

Juan and I went to see him.

Neither of us could stand him personally. He was totally ignorant, unsophisticated, and unpleasant. What he lacked in knowledge he made up with prejudice.

But he had connections. He knew people. And those people had money. Yet there was something else.[40] Something that would open Tar-Heel doors. Jordan Hill had associates in Acadiana (that is the Cajun part of the state) that had licensed proven technology and owned the rights to the patent. Ryan Richard and Michael Guidry[41] controlled Busan, a cyanide-based agriculture chemical compound that had been demonstrated on sewage sludge and other solid wastes. It was easy-to-use and safe. Hill claimed that he had witnessed Busan in action and had solicited investors. He wanted to tackle the pig waste problem. He had something that was proven and cost-effective.

Busan was a liquid that when mixed with sludge rendered it environmentally reliable and inert. There were no harmful pathogens and, most of all, no odors. Jordan Hill, though, was not a scientist. He needed *bona fide* researchers prepared to confirm that the chemical worked on pig waste. To be fair, the problem of pig poop is more complex than most would imagine. Odor control is mainly a subjective enterprise. Scientists around the world had been looking for a solution with limited success.[42]

Following the disaster in North Carolina, researchers around the world began to hasten their hog waste alternatives. One that seemed to gain traction among many environmental circles involved using constructed wetlands (for nutrient removal through the plants) combined with enzymes to eliminate pathogens.[43] The problem here, in hindsight, is obvious: the nutrient loading is too high for wetlands, and there are just too many pollutants for enzymes to eradicate.[44] Enzyme salesman popped up like ugly mushrooms everywhere. And everyone seemed to be pushing a different enzyme that would 'cure' the lagoon water and eliminate odors. For example, some used varieties of horseradish and peroxide.[45]

But the presence of hormones and pharmaceuticals in the waste rendered any enzyme less effective and wetlands all but useless.[46] In a massive CAFO study, researchers note "Endocrine disrupting compounds (EDCs) are chemicals that interfere with normal hormone function, either via synthetic chemicals that act or block normal hormonal activity or through exposure to high doses of natural hormones . . . EDCs are important because of increasing evidence linking exposure to reproductive and other health effects in humans and wildlife."[47] CAFOs were found to be a major source of EDCs harming aquatic ecologies. Wetlands and enzymes would not reduce EDCs in lagoons.

Another route that researchers took was to reevaluate lagoons. Lagoons operate on the principle of creating an anerobic zone (no oxygen) underneath the water so that denitrifying bacteria breakdown dangerous pollutants, like ammonia, into simpler compounds that can be then used safely on the farm.[48] Lagoons attempt to mimic nature. The problem is that nature does not stack hundreds of thousands of pigs in tiny cages and expect an open-pit to convert toxic waste into compost and nutrient-rich water.

Dr. Knecht and Josse looked around the country for solutions and funding. The natural place to start was the USDA or EPA. The problem was obvious: the UWMRC handled city problems, not rural ones. We were good with garbage, wastewater, and industrial pollution. Pigs lived (if you called it living) in the country. The USDA funded rural projects through land-grant, agricultural universities. The EPA was still looking at the issue.

Was this a city or country problem? Well, of course, the answer is both. City demand for pork had transformed the mom-and-pop farms into industrial CAFOs. The ecological impacts, though, as evidenced in North Carolina, and other hog states, hurt both communities. By the way, Mother Nature does not find our societal distinctions and political boundaries as relevant as we do. An interesting study confirmed that decades ago, but with one stipulation: rural people often are less aware of environmental issues and as such do not favor stronger regulations.[49] Remember: country people do not like government intrusion.

Neither the USDA nor EPA funded the proposals submitted by Al and Juan. Jordan Hill was disappointed and so were Richard and Guidry. I was still very busy with the final stages of my dissertation, and filming on a water documentary so I did not assist them with their grant applications. I do not think my collaboration would have made any difference. In a strange twist of fate, Hill was commiserating about the failure to get federal funding with his attorney Geno Henry.[50] Henry's reputation was that he fronted for organized crime and could not be trusted. He suffered with massive panic attacks, depression, and schizophrenia. He was another one of these lanky men well over six feet tall with an old, long horse-like face. He was about 70 years old when we met. The skin on his neck hung past his collar that seemed too tight. He talked incessantly as if he was giving closing arguments in his last case ever. Everything was terribly important to him. Before he died a few years later, Geno would snap, kidnap a small child, and barricade himself above a downtown bar. The police were able to rescue the child and Geno spent his remaining years hospitalized. He was mentally exhausted.

Geno was talking about his dilemma. He had an option to represent Richard and Guidry and their "wonder-cure Busan." But the clock was ticking. He needed to show progress in the next 90 days or his part in this play would be recast. In a popular breakfast restaurant in Lafayette, Louisiana, Geno ran into my father-in-law. Perhaps it was just "new son-in-law pride" he explained to Geno that I was the guy that could make this project move forward.

"We need to get Sutherlin involved," asserted Geno Henry to Jordan Hill.

"Why? I don't know any John Sutherlin. Who the hell is he?" he questioned.

Geno fired back, "The one to make this happen!"

Jordan thought about it. He then looked around to see what my reputation was. I did not know this until years later, but there is another John Sutherlin in Louisiana that is an environmental expert. His reputation was much better than mine. Maybe it still is. Hill confused the two of us and asked to see me. So, there I was sitting in his living room while he smoked, drank coffee, and cooked "red gravy" (spaghetti sauce). He used smoked pig feet to make the red gravy and it was delicious. Almost worth listening to him rant and rave.

Almost.

"We cannot just do the project," I explained, "We need money for supplies, testing equipment, laboratory analysis, and our time."

"How much?" Hill asked.

"It will take about six weeks, probably ten tests, equipment, independent lab analysis . . . so about $25,000." I offered unflinchingly. This was a few months before Hurricane Floyd wrecked the hog industry in North Carolina.

"Can you start now?" Hill demanded.

"Just cut the check."

I came back to the UWMRC and met with Juan Josse. I explained what happened and how I thought we should proceed. Being an engineer, Josse wanted a formal workplan and timetable. Again, he knew best, and we developed both. Then, we decided right then that we would split the money for our time. I was a political scientist, and he was an engineer. But we would both work on this project as equals. Rarely did he ever remind me that I was 'only' a political scientist.

The USDA, though, was right. They could never have funded our project. The UWMRC was housed at an urban university. We did not have any pig poop. I do not think there was any in all New Orleans. Even during Mardi Gras, it would be tough to find hog waste. But up the interstate was Louisiana State University (LSU). An agricultural research university. They had a facility called the Ben Hur Research Station (see Figure on the next page). And a Swine Unit.

And they had pigs. And lots of pigs meant that LSU had plenty of pig waste. They would be glad to 'donate' some to our research efforts. We were all just one big happy community of college professors and researchers. Why not? Juan and I bought two empty barrels with metal lids that were air-tight, a small cement mixer, several shovels, gloves, masks, and sampling equipment. Jordan Hill would provide the Busan. We were ready. Or so we thought.

By this time, Juan and I had worked in sewers, landfills, and industrial facilities that had compliance and odor concerns. We thought we understood stink. Hardly. The foul odor hits you in the face like a rubber mallet on a cold day. First, your nose. Then, your eyes. Eventually, you can taste the stench in your mouth. The ammonia stays in your clothes, hair, and car. And your lungs; causing massive respiratory damage.[51] A nasty cloud of urine and feces seemed to hover across the barns where the pigs were kept. There were only about 300 head here. I could not imagine how workers or neighbors could be

Figure 2.1. LSU Ben Hur Research Stations, where project began. Courtesy of the Author.

around hundreds of thousands of smelly, squealing pigs. Country air should be fresh and pure. Exhaust fumes in a city or the smell of a garbage truck can be a nuisance. This was intolerable. I cannot say that I ever got used to the smell.

By noon, the summer sun had burned off the hog haze and we had filled up our two 55-gallon barrels with the most concentrated pig waste we could get. Nothing from the lagoon. This was from under the slats in the floor where urine and feces dropped. It was unfiltered, untreated, and undiluted. This was pure, unadulterated pig crap. We rolled the barrels into the back of Juan's truck, shut the gate, and proceeded back to New Orleans. I am not sure, but I can guess that he was thinking the same thing as me: "What in the hell have we gotten ourselves into?"

The barrels were stored in an outside, but covered area, and placed under lock. We posted signs on the gate and went to the lab to set up for the following day's experiments. Jordan Hill had given us plenty of Busan to complete the testing. Early the next morning, Juan and I covered our mouths and faces with protective gear, mixed the Busan and pig waste, and stirred the blend in the cement mixer. Since we wanted to see how effective Busan was, we had various times for mixing. The goal here was to use as less as possible to reduce as much as possible. We were trying to find the right threshold for the product to be cost-effective.

The initial trials were not successful. There was no measurable reduction in ammonia or any pathogen. We clearly had done something wrong. So, we repeated the experiments for several more days. After all, we had more than 100 gallons of pig waste and all the Busan we needed. Yet nothing seemed to work. Nothing. No changes. The ammonia levels never dropped, the bacteria count stayed the same, and it smelt like hell.

I called Jordan Hill.

"The results are not what we were expecting," I explained.

"What are you doing wrong?" he asked.

"Well, I am not sure. Do you have the patent instructions?"

"Of course."

"Send them to me. Or I will come get them tonight or tomorrow."

"Ok, don't let me down. I got a lot riding on this. I have also talked some of my friends and two widows into investing." He was not joking.

We got the instructions.

Immediately we smelled more than pig poop.

Here is a summary of the instructions: mix 1 part sludge with six parts sand, 6 parts peroxide, 6 parts water, six parts Busan. Then expect to find an 87% reduction.

What?

Juan and I were in shock. The dilution caused by sand and water meant the waste would have less pollutants simply due to the concentration. Adding peroxide had some benefit. But Busan added nothing to the process. In fact, Busan was just a brand name for commercial grade cyanide. It killed pathogens. Sure. But was this really the safe, environmentally friendly alternative that we were "confirming"? Not likely.

But it did not stop there. Busan came in a powered form. What we had been given was a liquid. Jordan talked about using it and seeing the results.

I called Jordan Hill right away.

"Jordan, when you saw it used to reduce odors and pathogens in sewage sludge, what was the process? I inquired.

"I guess Richard and Guidry poured some on the sludge, stirred it, and we left it covered and sitting in the lab. We left for lunch and when we came back it was 'odor-free.'"

"Are you sure you saw it poured into the sludge," I pressed him hard on this point.

"100 percent!" He exclaimed.

"Jordan, what you saw was not Busan. I am not sure what you saw, but it was not Busan. It was probably some liquid odor reducer, but it was not Busan."

The next sound I heard sounded like someone tripping over a table at night. Like the sound of someone going to the bathroom in a hurry. Then, a loud thump. Jordan had fainted. He caught himself on the side of the table next to his chair but managed to knock everything off and onto the floor. He was probably on his knees when he regained his composure.

"What am I going to do? Those bastards Richard and Guidry hustled me. Lied to me. I lied to my friends. I got friends to invest their lifesavings. I made promises."

I offered some hope, "Let Juan and I think about this and get back to you."

"Please don't say anything to Geno."

We had no reason to talk with Geno Henry. He talked enough for everyone already.

Juan and I met the following day to figure things out.

Neither of us knew what would happen next.

NOTES

1. National Weather Service. Hurricane Floyd storm summary. Located at https://www.weather.gov/mhx/Sep161999EventReview. Accessed 12 May 2022.

2. Ibid.

3. Moore, R., & Barnes, J. (2004). *Faces from the flood: Hurricane Floyd remembered.* UNC Press Books. The authors donated all proceeds from the book to the American Red Cross.

4. Ibid.

5. Thompson, C. D., & Amberg, R. (2001). The Great Deluge: A Chronicle of the Aftermath of Hurricane Floyd. *Southern Cultures, 7*(3), 65–82.

6. Tabachow, R. M., Peirce, J. J., & Essiger, C. (2001). Hurricane-loaded soil: Effects on nitric oxide emissions from soil. *Journal of Environmental Quality, 30*(6), 1904–1910.

7. Campese, C. (2000). Preparation, experience, and aftermath of hurricane Floyd. *AORN Journal, 72(*1), 82–93.

8. Setzer, C., & Domino, M. E. (2004). Medicaid outpatient utilization for waterborne pathogenic illness following Hurricane Floyd. *Public Health Reports, 119*(5), 472–478.

9. Schmidt, C. W. (2000). Lessons from the flood: will Floyd change livestock farming? *Environmental Health Perspectives, 108*(2), A74–A77.

10. Letson, D., Gollehon, N., Kascak, C., Breneman, V., & Mose, C. (1998). Confined animal production and groundwater protection. *Applied Economic Perspectives and Policy, 20*(2), 348–364.

11. Federal Agriculture Improvement and Reform Act of 1996. 7 USC 7201 et seq. Located at https://www.congress.gov/104/plaws/publ127/PLAW-104publ127.pdf. Accessed 15 May 2022.

12. Letson et al.

13. Fleming, R. A., & Long, J. D. (2002). Measuring the cost of restricting access to cropland for manure nutrient management. *Agronomy Journal, 94*(1), 57–64.

14. This is an interesting point considering climate change concerns along the coast. Should CAFOs have to move or be banned as part of an adaptation strategy? Researchers have asked that recently. Stoddard, E. A., & Hovorka, A. (2019). Animals, vulnerability, and global environmental change: The case of farmed pigs in concentrated animal feeding operations in North Carolina. *Geoforum, 100*, 153–165.

15. Lobban, R. A. (1999). The Perils of Industrial Pig Farming. The author notes that other animals lost their lives that day: "2.4 million dead chickens . . . [and] 500,000 dead turkeys."

16. Ibid.

17. Morris and Barnes.

18. Harmin, C. (2015). Flood vulnerability of hog farms in Eastern North Carolina: An inconvenient poop. (Master's Thesis, East Carolina University). Retrieved from the Scholarship. (http://hdl.handle.net/10342/5143.)

19. Tietz, J. (2006). Boss hog. *Rolling Stone, 1015*, 89–90.

20. Smithfield Foods website. Located at https://www.smithfieldfoods.com/About -Us#history. Accessed 10 June 2022.

21. Tiez. The author adds, "Hogs produce three times more excrement than human beings do. The 500,000 pigs at a single Smithfield subsidiary in Utah generate more fecal matter each year than the 1.5 million inhabitants of Manhattan. The best estimates put Smithfield's total waste discharge at 26 million tons a year. That would fill four Yankee Stadiums. Even when divided among the many small pig production units that surround the company's slaughterhouses, that is not a containable amount."

22. Ladd, A. E., & Edward, B. (2002). Corporate swine and capitalist pigs: A decade of environmental injustice and protest in North Carolina. *Social Justice, 29*(3), 26–46.

23. We Are Democracy. Hog Money Pollutes General Assembly. July 13, 1995. Located at https://democracync.org/research/july-1995-hog-money-pollutes-nc -general-assembly/. Accessed 11 June 2022.

24. Fine, Ken and Erica Hellerstein. Big pork has given $272,000 to house republicans who voted in favor of hog-farm-protection bill. Indy Week. April 7, 2017. Located at https://indyweek.com/news/archives/big-pork-given-272-000-house -republicans-voted-favor-hog-farm-protection-bill/. Accessed 10 June 2022.

25. North Carolina Assembly Bill 467. Located athttps://www.ncleg.net/Sessions /2017/Bills/House/PDF/H467v1.pdf. Accessed 11 June 2022.

26. Hellerstein, Erica. Bill Would protect hog farmers from lawsuits; activists and neighbors aren't pleased. Indy Week. March 30, 2017. Located at https://indyweek .com/news/archives/bill-protect-hog-farmers-lawsuits-activists-neighbors-pleased/. Accessed 11 June 2022.

27. Ibid.

28. Savan, B., Morgan, A. J., & Gore, C. (2003). Volunteer environmental monitoring and the role of the universities: the case of Citizens' Environment Watch. *Environmental Management, 31*(5), 0561–0568.

29. Hellerstein.

30. Lopez, R. A. (2001). Campaign contributions and agricultural subsidies. *Economics & Politics, 13*(3), 257–279.

31. Stratmann, T. (1991). What do campaign contributions buy? Deciphering causal effects of money and votes. *Southern Economic Journal*, 606–620.

32. Chappell, H. W. (1982). Campaign contributions and congressional voting: A simultaneous probit-tobit model. *The Review of Economics and Statistics*, 77–83.

33. Kozak, Catherine. Hogs after Floyd: Nothing's changed. September 18, 2014. CoastalReview.Org. Located at https://coastalreview.org/2014/09/hogs-after -floyd-nothings-changed/. Accessed 11 June 2022.

34. Applewhite, J. (2000). Winging the flood. *Carolina Quarterly, 52*(3), 20.

35. EPA Research Centers. Located at https://cfpub.epa.gov/ncer_abstracts/ index.cfm/fuseaction/outlinks.centers/center/73/format/Tab. Accessed 2 April 2022. According to the EPA, the UWMRC's major goal was to "provide an integrated multimedia waste management approach to the solution of urban environmental problems, and to advance the state of the art of urban waste management and pollution prevention through technology transfer."

36. Times-Picayune. Obituary. Albert Knecht. May 2, 2019. Located at https://obits .nola.com/us/obituaries/nola/name/albert-knecht-obituary?id=1830508. Accessed 12 February 2022.

37. Causeway Bridge. Located at https://www.thecauseway.us/. Accessed 10 February 2022.

38. Salt Lake City Tribune. Obituary. Robert S. Jordan. April 29, 2022. Located at https://www.legacy.com/us/obituaries/saltlaketribune/name/robert-jordan-obituary?id =34728416. Accessed 4 June 2022.

39. His name has been changed to protect his family's privacy.

40. Maurer, D. (2010). *The big con: The story of the confidence man.* Anchor. The con artist is often being conned by others. I do not think Hill thought of himself as a hustler.

41. Their names have been changed to protect their family's privacy.

42. Aneja, V. P., Chauhan, J. P., & Walker, J. T. (2000). Characterization of atmospheric ammonia emissions from swine waste storage and treatment lagoons. *Journal of Geophysical Research: Atmospheres, 105*(D9), 11535–11545.

43. Hunt, P. G., & Poach, M. E. (2001). State of the art for animal wastewater treatment in constructed wetlands. *Water Science and Technology, 44*(11–12), 19–25.

44. Baddam, R., Reddy, G. B., Raczkowski, C., & Cyrus, J. S. (2016). Activity of soil enzymes in constructed wetlands treated with swine wastewater. *Ecological Engineering, 91*, 24–30.

45. Eniola, B., Perschbacher-Buser, Z., Caraway, E., Ghosh, N., Olsen, M., & Parker, D. (2006). Odor control in waste management lagoons via reduction of p-cresol using horseradish peroxidase. In 2006 ASAE Annual Meeting (p. 1). American Society of Agricultural and Biological Engineers.

46. Pruden, A. (2009). *Hormones and pharmaceuticals generated by concentrated animal feeding operations.* L. S. Shore (Ed.). Springer New York.

47. Easton, J. H., Dongell, A. R., & Oberdörster, E. (2005). Instream steroid hormone levels in a CAFO-impacted watershed and biological removal strategies. In *Impacts of Global Climate Change* (pp. 1–12).

48. Hunt, P. G., Matheny, T. A., Ro, K. S., Vanotti, M. B., & Ducey, T. F. (2010). Denitrification in anaerobic lagoons used to treat swine wastewater. *Journal of Environmental Quality, 39*(5), 1821–1828.

49. Tremblay, K. R., & Dunlap, R. E. (1978). Rural-urban residence and concern with environmental quality: A replication and extension. *Rural Sociology, 43*(3), 474.

50. His name has been changed to protect his family's privacy.

51. Schultz, A. A., Peppard, P., Gangnon, R. E., & Malecki, K. M. (2019). Residential proximity to concentrated animal feeding operations and allergic and respiratory disease. *Environment International, 130*, 104911.

Chapter 3

Basketball, Hustlers, and Pigs

I love basketball. The game is exciting. On any given night five guys can get hot and beat any other team. Jordan Hill and Geno Henry were basketball players. As in, they played basketball in college. They knew the players, the game, and the rules. When their beloved Busan was discovered to be a fraud perpetrated by Richard and Guidry, it was like they had gotten double technical fouls in the state championship game and been asked to leave the arena. They had no chance at being in the game. Desperation set in. Both knew the clock was ticking and they were running out of options. Juan and I were sent in off the bench to save the game. Just a few weeks earlier, to continue with this awful basketball metaphor, we barely knew there was a game, hardly could play, and did not have a uniform or shoes.

Yet here we were.

Poised to step onto the court, make a pass, take a charge, and, hopefully, make some free-throws.

And why not?

Was the problem more complex than any other environmental issue we had tackled?

It certainly was not more complex than solid waste management. Imagine the multifaceted aspects of trash: household generation, maybe some separation, collection at the curb, transport by trucks, and disposal at the landfill. A hundred things can go wrong.[1] And sewage with its network of pipes collecting wastewater *enroute* to a treatment plant. Pipes burst or people shovel through them.[2] What people flush down their toilets is disturbing and hardly encourages them to consider their ecological impacts.[3] Sewage in people's yards or backed-up in their toilets guarantees that citizens will call City Hall. Polluted properties called Brownfields have a tangled legal and regulatory history together with an assortment of contaminants dispersed in buildings, equipment, and land.[4]

These were hard environmental issues to untangle.

Sorting out what is in pig waste is not all that hard. The composition of chemicals, nutrients, and metals had been known for years.[5] Sure, variables such as ambient temperature, pH, and rainfall impact the waste content and ultimately the options for treatment.[6] However, there is considerable variability in pig waste depending on their age, diet, and how often they are washed down.[7] The source of the wash down water (i.e., pond, potable, recycled) can have a massive impact of the chemical oxygen demand (COD) loading and even the smell.[8] Also, organic matter decomposes during the anerobic phase of "treatment" and that changed the moisture and nutrient value.[9]

As such, there have been lots of studies analyzing the impact of adding additional carbon sources to stabilize the decomposition and reduce the water-content of pig waste. Readily available sources include other manures (i.e., cow),[10] sawdust,[11] and municipal waste from recycling (i.e., paper, ground-up limbs, or yard waste) and compost operations.[12] An interesting "natural" alternative included planting flowers, gum trees (eucalyptus), and water hyacinths to uptake nutrients and water.[13] I have seen hundreds of "ideas" on how to "solve" this issue. Very few have any merit whatsoever.

The science behind lagoons often leads farmers and researchers to attempt to short-circuit the biological and chemical demands needed for proper waste management. When operated properly, within the limits of volume (both water and solids), number of pigs, and spraying fields, a lagoon could be a viable part of the farm operations. CAFOs changed that by putting way too many pigs in one central location. The American Society of Agricultural and Biological Engineers made this point very clear when analyzing concerns regarding lagoons: "Two challenges must be addressed if lagoons are to remain a viable treatment alternative for animal agriculture: inefficient recovery of plant nutrients, and odor and ammonia emissions."[14] These simple concepts demand sophisticated science skills. Lagoon science is not just about digging a hole and filling it in with urine and feces.

In many ways, the simplicity of lagoons lured us into thinking of an equally simple response. This was an early mistake. One that was replicated by scientists all over the world.

Testing pig waste was also a phenomenally technical issue. You do not just go out to the lagoon and grab a sample and send off to the laboratory for analysis. Here, there is every imaginable device and technique. One study took 122 days of sampling across dozens of parameters.[15] Even with only five variables of chemicals and nutrients, adding water-content, temperature, and pH could result in 1,000s of tests. Even a more recent study (2005), examined whether a mobile spectroscopy instrument would be feasible.[16] The industry was attempting to improve its practical knowledge using the latest technology. Building on that research, waste from the floor of the cages was analyzed to determine the fat content of pig waste.[17] The goal there was to find a quicker, more reliable method besides waiting for a laboratory to complete their testing. Having access to data in the field would be critical.

But that was still years away.

We were in the late 1990s.

The solid waste portion was just one aspect of a total solution. What about the gas created when the solids fraction broke down? The biogas. Biogas is the methane fraction (CH_4) in the air emissions that can be captured and turned into energy. CAFO lagoons are a significant source of biogas (and methane).[18] At some CAFOs, the lagoons are under the slats in the floor where the pigs stand, eat, and excrete waste. The result is massive amounts of odors and methane gas generation.[19] Few would argue that this is not a problem. But, what to do? Cost must be factored into the equation. Ignoring the financial side of any solution will doom it before it can start. Again, this was a *Clean Water Act* (cost-benefit) and not a *Clean Air Act* (health-based) issue for the USDA and the EPA. Some people suggested that this could even be a better, low-cost approach to managing this issue.[20] Then, at an international conference in Iowa, it was demonstrated that these methods were at least competitive with standard lagoon technology.[21]

Maybe there was a light shining on the lagoon?

International researchers had been more active here. Asian universities, as was the case for solids waste, had conducted research because of China and Japan being major consumers, and, in the case of China, massive producers. They needed an alternative.[22] China has been a leader in biogas research for many years.[23] But with so much land, cheap labor, and few environmental restraints, China offered little to the West.[24] So, Europeans began to look for answers on their own based in Asian research. Biodigesters that captured the air emissions from decomposed solids was one such technology.[25] The Netherlands had led the way here for years. In the early 1970s, the Dutch had studied technologies and methods for combining manure from farms, sludge from sewage plants, and parts of the waste from landfills.[26]

American researchers conducted a comparative analysis of various digester technologies.[27] To be fair, Pennsylvania researchers had considered anerobic digesters in the 1970s, but this was for dairy cattle.[28] Cow manure is different than pig poop. While they both contain significant amounts of nitrogen (N), phosphorus (P), and potassium (K), the management and housing of dairy cows (often outside on the range) is vastly different than the tight, cramped quarters for swine.[29] Cattle bred for slaughter, though, have just as wretched a life as hogs.

By the way, the NPK compound has been known to be an excellent fertilizer for decades.[30] Perhaps this could be a value-added benefit to any solution? An "organic" or non-chemical fertilizer as the by-product of waste management? Maybe. Back to understanding cow manure. As one report noted, "Dairy cattle often spend portions of their time in pasture areas, feeding and lounging barns, and milking parlors. From an environmental standpoint, manure dropped in any of these locations may be of concern."[31]

Also, having been to both dairy farms and pig CAFOs, their diets are very different. Cows eat hay; pigs eat grain. That means their feces is different. Cow manure is fibrous and does not smell as bad (less ammonia). In other words, technologies and management changes applicable to cows would not simply transfer to pigs. Chicken manure is even harder to manage.[32] I had to compost chicken waste when I got certified as a Compost Facility Operator. It is very wet, pasty, and the ammonia levels are off the charts.[33] But doing a massive literature review cow and chicken manures would be key for our future work.[34]

So back to the pigs.

Yes, of course, their waste was different. There was more of it per volume and per pound of pig produced and it smelt worse.

Maybe there was some hope here.

If biodigesters could handle some of the solids fraction and the air emissions, then Juan and I would be free to focus on cleaning up the water. He was a wastewater expert. I was trying to simply keep my head above the waves. Yet, the Dutch solution was not adequate. There is considerable variability between CAFOs. Producing a reliable biogas for the purposes of fuel generation from waste streams has been the goal of every sustainable, environmental plan ever.[35] Yet there were concerns. The composition of the pig waste impacted the rate of biogas production and the moisture content, which directly impacts the BTU value of the biogas.[36] Ambient temperatures due to changes in the seasons directly altered the BTU value. Studies in several countries, Pakistan, China, and France, further demonstrated the limits to reliance on digesters.[37] Once again, the Netherlands saw a plethora of projects attempting to improve the efficiency of digesters by manipulating internal temperatures.[38]

There was even research showing that constantly stirring the swine manure increased biogas production.[39] American research even borrowed some of these ideas and digesters were all the rage across many waste streams, especially from CAFOs, during the 1990s.[40] Could this be a low-tech alternative? The potential for reducing air emissions (and odors), eliminating lagoons, and producing energy from pig waste was very attractive.[41]

What we learned was straightforward: mixing other waste streams with pig waste typically lowered the BTU value of the biogas, temperature mattered, and reducing the moisture was essential. Also, any technology required that the waste be consistent. Which is why mixing sewage sludge, and household trash, or other animals wastes made things more problematic. It was critical that all the waste be pre-processed, say, through a grinder to make it consistent. A homogenous slurry from the lagoon, though, meant the solids sunk to the bottom and the liquid rose to the top while the methane gas bubbled up and escaped as fugitive emissions.

In other words, pig waste was much more intricate of an issue than fixing a lagoon filled with poop. We did not have 122 days to find a solution. We had less than a month to come up with a research strategy that did not include Busan or any other enzyme-magic powder or *gris gris*. Anything we did have had to have real science backing it up. We worked for a university through an EPA-funded environmental think tank. No gimmicks. No tricks.

Jordan Hill met with us to discuss how to proceed.

"Can you do it?"

Given any amount of money and time, most environmental issues can be "resolved." The question was not finding a solution. The solution was banning CAFOs for pigs (and maybe most other animals).[42] That would have solved the issue almost instantly. Why have CAFOs in the first place? When Texas began expanding hog production in the panhandle area, this was a central question.[43] But states like Iowa and North Carolina were where the pigs were. Texas was a small player. Yet, following Hurricane Floyd, many began questioning the sanity of allowing such high concentrations of pigs in CAFOs in flood zones. Maybe allowing local governments to enact zoning ordinances for CAFOs could help.[44] Clearly, the EPA and all its regulations from the 1970s were never designed to address water and air pollution from CAFOs.[45] The EPA issues discharge permits (National Pollutant Discharge and Elimination System under the CWA) for "point-source" like the end of pipes. Lagoons are diffused and not a single source. CAFO reform was unlikely then. Still, the need for such reform remains.[46]

I thought, and then responded, "We can do it."

Hill exhaled a thick plume of cigarette smoke and swallowed some black coffee.

"Thanks."

I explained how we would put together the proposal and he would need to get the money. Not $25,000 for a laboratory analysis. This would be much more. Ten times as much.

"I cannot see how we can do this in under a year and for less than $225,000," I said.

Juan sat quietly. Probably he was in disbelief. He did not flinch. He knew that I had found that number from the nether world of budget making. That is a nice way of saying I pulled it out of my ass.

"That is a lot of money. I cannot go back to my friends for that," Hill remarked.

"Ok, then where?" I questioned, then added, "And not that Geno Henry guy either. Something is not right with him." I questioned.

He took a long drag off another cigarette, then avowed, "Drexel Smith."[47]
I repeated the name over and over.

Drexel Smith.

"Yeah. We've known each other for years. Basketball is a small world," Jordan Hill affirmed.

And he did. Among many other basketball players that were household names in the 1970s.

First, though, Jordan wanted to reach out to a friend that knew investors from the northeast.

Within a couple of weeks, Jordan delivered a check that would "get us started." The check was drawn on a Boston bank where the investor was a Vice-President. Juan and I met with our Urban Waste Management & Research (UWMRC) Director. We outlined the project and gave him the check. Juan and I began ordering tools, piping, tanks, and protective gear. Within three days, we were informed the check had bounced.

It was no good. Rubber!

How? How can someone write a bad check from an account in which they are one of the senior managers of the bank? Not possible. Yet it happened. We never got a full explanation of what exactly happened.

Now, we were back to square one.

Jordan Hill with few other options reached out to one of his closest basketball buddies. There was a handful of telephone conversations and a back and forth of proposals and budgets. We finally agreed to $225,000 over 18 months. Starting immediately. The payments would be made in four installments with half being paid up front. There would be tons of equipment needed to get this "project" going. Juan and I had already made purchases and needed to get reimbursed.

"This is the second largest check I have ever written," explained Hill.

"I'll try not to lose it," I quipped.

"Do you really think you boys can do this?" He pleaded once again.

"I sure hope so," I said. No expression. No sales-pitch. We were in uncharted waters. Up the proverbial creek without a paddle. Lots of smarter, better funded researchers around the globe had tried . . . and failed.

Juan and I left there knowing the immense pressure we had placed ourselves.

The first thing we needed was a steady supply of pig waste. Actual farms were not an option due to access, location, and potential costs. We had developed a relationship with the LSU Ben Hur Station in Baton Rouge. I was still newly married and was living in the middle of nowhere at this point. It made sense for me to move to Baton Rouge to be close to the facility. They had a Swine Unit (see Figure 3.1). And Juan wanted to stay in New Orleans. At the time, I was also teaching at Tulane University. Commuting to New Orleans

Figure 3.1. Swine Unit, where the waste was collected, tested, and treated. Courtesy of the Author.

from Baton Rouge was just as long. And, because my wife was pregnant, I preferred to live closer to a hospital anyway. Baton Rouge had plenty of options. There may have been a mid-wife and a Voodoo Priestess where we were living. I know there was a registered sex-offender and a cock-fighter (the same person). He was required by law to inform us of the former. I found out about the latter crime upon looking for his house.

I had no arguments when I suggested that we move.

We found a small place close to the Swine Unit at the Ben Hur Station and moved.

This place really was our first home. We had lived elsewhere, but that place in Baton Rouge was a cute, pink house with bay windows and a well-kept lawn. A house. Two cars. Wife and a child on the way. I felt very middle class for the first time in my life. I even bought a grill and talked to the neighbors. Sometimes I even worried about my yard.

Happiness is often best as a reflective emotion. Many times, I do not think we realize how happy we were in a situation until it is over. And too often we do not even know it until we face a dreadful moment. At that time, I was teaching and doing research. It was everything I had wanted to do.

Perhaps, the reason I remember the place with so much nostalgia is because there was so much potential. Possibilities seemed endless. Newly married, new job, new child, new house . . . But there was also so much personal tragedy lurking in the shadows. Within a few weeks of getting this project started, my mother was diagnosed with stage-4 endometrial cancer. It had spread everywhere. She had been given less than a year to live. She died within 6 months.

My mother died without seeing me complete my PhD or see her first grandchild born.

This has bothered me my entire life. My mother was my pal. A good friend. Someone that I could always rely upon. No one loves the way a mother loves.

I did what I could to move forward.

Work is often the best cure for sorrow. So, I worked on the farm.

Juan and I spent all day (and sometime the night) at the Swine Unit, bathed, rested, ate something (not bacon or pork chops) and hit the computer to do research. Lots of research. We knew that we still had much to learn.

To make sure it is clear: the failed Busan lab analysis was first; then, we got full funding for the Swine Unit project, and then Hurricane Floyd hit North Carolina. We were in the midst of doing pig waste research at the optimal time.

Fate. Karma. Clean-living.

Whatever it was, Juan and I were becoming experts.

And, within a year, we would file our first patent.

The Swine Unit was more than happy to give us all the crap we wanted.

If we stayed out of their way, did not disturb the pigs, and did not build anything permanent, then everything would be fine. Also, we could never ask LSU workers for any help. We paid them to give us poop. That was it. So, we rented a flatbed trailer that became the focus on our farm research and put it next to the pig house and got to work (see Figure 3.2). The trailer idea was "forced" on us due to space limitations. We tried our best to not interfere with the other farm-related research projects that were going on at the Swine Unit.

Juan and I even hired a couple of workers: Don and Wayne.

They built structures needed to work on the trailer. They built railings, stairs, and helped with all piping work. We never asked them to do anything too dangerous but there was an inherit danger in everything we did. It was hot and it smelt bad, and the waste always found a way to cling to your clothes, hair, or in your nose. Beyond the obvious odor concerns, we were isolated, and any required medical help would have been too late to provide any assistance.

Every day presented us with an opportunity to honorably quit and do some quiet research from the comfort of our university offices. There were colleagues working in air-conditioned offices and analyzing data. We could have been writing articles and giving lectures.

We chose to continue working at the Swine Unit.

Figure 3.2. The Research Trailer, where the invention was built. Courtesy of the Author.

Rain or shine, hot or cold, Juan and I labored on that trailer. Finding ways to economically place tanks, hook up pumps, and values to treat pig slurry. The forced space always kept us from going too far into other areas of the Swine Unit. It also kept our costs lower.

We were now neck deep in the . . . lagoon!

NOTES

1. Heyer, K. U., & Stegmann, R. (2005). Landfill systems, sanitary landfilling of solid wastes, and long-term problems with leachate. *Environmental Biotechnology*, 375–394.

2. Mattsson, J., Hedström, A., Ashley, R. M., & Viklander, M. (2015). Impacts and managerial implications for sewer systems due to recent changes to inputs in domestic wastewater–A review. *Journal of Environmental Management, 161*, 188–197.

3. Narain, S. (2002). The flush toilet is ecologically mindless. *Down to Earth, 10*(19), 1–14.

4. Sarni, W. (2010). *Greening brownfields: Remediation through sustainable development.* McGraw-Hill Education.

5. Hammond, E. G., Heppner, C., & Smith, R. (1989). Odors of swine waste lagoons. *Agriculture, Ecosystems & Environment, 25*(2–3), 103–110.

6. Cullimore, D. R., Maule, A., & Mansuy, N. (1985). Ambient temperature methanogenesis from pig manure waste lagoons: Thermal gradient incubator studies. *Agricultural Wastes, 12*(2), 147–157.

7. Tiquia, S. M., Tam, N. F. Y., & Hodgkiss, I. J. (1998). Changes in chemical properties during composting of spent pig litter at different moisture contents. *Agriculture, Ecosystems & Environment, 67*(1), 79–89.

8. Ros, M., Garcia, C., & Hernández, T. (2006). A full-scale study of treatment of pig slurry by composting: Kinetic changes in chemical and microbial properties. *Waste Management, 26*(10), 1108–1118.

9. Hsu, J. H., & Lo, S. L. (1999). Chemical and spectroscopic analysis of organic matter transformations during composting of pig manure. *Environmental Pollution, 104*(2), 189–196.

10. Inbar, Y., Chen, Y., & Hadar, Y. (1990). Humic substances formed during the composting of organic matter. *Soil Science Society of America Journal, 54*(5), 1316–1323.

11. Huang, G. F., Wu, Q. T., Wong, J. W. C., & Nagar, B. B. (2006). Transformation of organic matter during co-composting of pig manure with sawdust. *Bioresource Technology, 97*(15), 1834–1842.

12. Chefetz, B., Hatcher, P. G., Hadar, Y., & Chen, Y. (1996). Chemical and biological characterization of organic matter during composting of municipal solid waste (Vol. 25, No. 4, pp. 776–785). American Society of Agronomy, Crop Science Society of America, and Soil Science Society of America.

13. Mbendo, J., & Thomas, T. H. (1988). Economic utilization of water hyacinth from Lake Victoria. *Development Technology Unit Working Paper* (51).

14. Hamilton, D. W., Fathepure, B., Fulhage, C. D., Clarkson, W., & Lalman, J. (2006). *Treatment lagoons for animal agriculture.* The authors added, "Up to 80% of all nitrogen entering lagoons cannot be accounted for in lagoon effluent, and a great portion of manure phosphorus entering lagoons is retained in sludge. Plant nutrients are less concentrated in lagoon effluent than in other manure treatment products, although lagoon effluent has a better balance of nitrogen to soluble phosphorus than most sources of manure nutrients. Lagoon effluent should be used in crop production on a nitrogen basis, irrigating effluent in multiple applications throughout the growing season. Managing effluent in this manner requires expensive, permanent irrigation equipment to apply what is essentially low-quality fertilizer. Nitrogen application is inherently out of sync with phosphorus since the majority of manure phosphorus is only recovered when solids are removed at the end of the sludge storage cycle, which may last as long as 10 to 20 years."

15. Hsu et al.

16. Saeys, W., Mouazen, A. M., & Ramon, H. (2005). Potential for onsite and online analysis of pig manure using visible and near infrared reflectance spectroscopy. *Biosystems Engineering, 91*(4), 393–402.

17. Yang, Z., Han, L., & Fan, X. (2006). Rapidly estimating nutrient contents of fattening pig manure from floor scrapings by near infrared reflectance spectroscopy. *Journal of Near Infrared Spectroscopy, 14*(4), 261–268.

18. Sharpe, R. R., & Harper, L. A. (1999). Methane emissions from an anaerobic swine lagoon. *Atmospheric Environment, 33*(22), 3627–3633.

19. Morgenroth, E. (2000, November). Opportunities for nutrient recovery in handling of animal residuals. In Animal Residuals Management Conference, Kansas City, Missouri (pp. 1–10).

20. Moser, M. (1998). A Low-Cost Digester to Control Odors at a 120,000 Head Hog Farm. In 2001 ASAE Annual Meeting (p. 1). American Society of Agricultural and Biological Engineers.

21. Moser, M. A. (1998). Anaerobic digesters control odors, reduce pathogens, improve nutrient manageability, can be cost competitive with lagoons, and provide energy too. Resource Conservation Management, Inc. Presentation at Iowa State University.

22. Chin, K. K., & Ong, S. L. (1993). A wastewater treatment system for an industrialized pig farm. *Water Science and Technology, 28*(7), 217–222.

23. Nianguo, L. (1984). Biogas in China. *Trends in Biotechnology, 2*(3), 77–79.

24. Geping, Q. (1982). Environmental protection in China. *Environmental Conservation, 9(*1), 31–33. One of my favorite works here is Shapiro, J. (2001). *Mao's war against nature: Politics and the environment in revolutionary China.* Cambridge University Press.

25. An, B. X., & Preston, T. R. (1999). Gas production from pig manure fed at different loading rates to polyethylene tubular biodigesters. *Livestock Research for Rural Development, 11*(1), 1–8.

26. Raven, R., & Verbong, G. (2004). Dung, sludge, and landfill: Biogas technology in the Netherlands, 1970–2000. *Technology and Culture, 45*(3), 519–539.

27. Wright, P., Inglis, S., Ma, J., Gooch, C., Aldrich, B., Meister, A., & Scott, N. (2004). Comparison of five anaerobic digestion systems on dairy farms. In 2004 ASAE Annual Meeting (p. 1). American Society of Agricultural and Biological Engineers.

28. Persson, S. P., Bartlett, H. D., Branding, A. E., & Regan, R. W. (1979). Agricultural anaerobic digesters: design and operation (No. NP-2901472). Pennsylvania State Univ., University Park (USA). Agricultural Experiment Station.

29. Midwest Plan Service. 1975a. Livestock waste facilities handbook. MWPS–18. Iowa State University, Ames, IA.

30. Prummel, J. (1960). Placement of a compound (NPK) fertilizer compared with straight fertilizers. *Netherlands Journal of Agricultural Science, 8*(2), 149–154.

31. Hubbard, R. K., Lowrance, R. R., & Wright, R. J. (1998). *Management of dairy cattle manure.* Agric. Uses Munic. Anim. Ind. Byprod, 91–102. The authors did note, "However, unless too many cattle are pastured per area of land or unless cattle are allowed free access to streams, lakes, or ponds, manure dropped in pasture areas may be of less environmental concern than that in barns and milking areas. Manure dropped by cattle while in the feeding and lounging barns and milking parlor is in effect a point source of nutrients that must be used. Point sources of pollution include such things as chemical spills, septic tanks, and so forth, and the manure dropped in barns and parlors is a point source in the sense that the land area where it is dropped

does not have the capacity to filter the load. Water added from cleaning of tanks or utensils in the milk house also contributes to the total amount of manure load."

32. Bujoczek, G., Oleszkiewicz, J., Sparling, R. R. C. S., & Cenkowski, S. (2000). High solid anaerobic digestion of chicken manure. *Journal of Agricultural Engineering Research, 76*(1), 51–60.

33. Elwell, D. L., Keener, H. M., Carey, D. S., & Schlak, P. P. (1998). Composting unamended chicken manure. *Compost Science & Utilization, 6*(2), 22–35.

34. Indriyati, L. T., & Goto, I. (1997). Effect of zeolite addition to chicken manure on nitrogen mineralization in the soil. In *Plant Nutrition for Sustainable Food Production and Environment* (pp. 593–594). Springer, Dordrecht. This is where the idea for using Zeolite originated. But that is much later.

35. Marchaim, U. (1992). *Biogas processes for sustainable development* (No. 95–96). Food & Agriculture Organization.

36. Sambo, A. S., Garba, B., & Danshehu, B. G. (1995). Effect of some operating parameters on biogas production rate. *Renewable Energy, 6*(3), 343–344.

37. Nazir, M. (1991). Biogas plants construction technology for rural areas. *Bioresource Technology, 35*(3), 283–289.

38. Ahring, B. K. (1995). Methanogenesis in thermophilic biogas reactors. *Antonie van Leeuwenhoek, 67*(1), 91–102.

39. Angelidaki, I., & Ahring, B. K. (2000). Methods for increasing the biogas potential from the recalcitrant organic matter contained in manure. *Water Science and Technology, 41*(3), 189–194.

40. Gettier, S. W., & Roberts, M. (1994). Swine lagoon biogas utilization system (No. CONF-9410176-). Western Regional Biomass Energy Program, Reno, NV (United States).

41. Van Dyne, D. L., & Weber, J. A. (1994). Biogas production from animal manures: What is the potential? Industrial Uses/IUS-4/Special Article.

42. During the COVID pandemic, this issue has arisen (again). Could better treatment for animals prevented the outbreak? Some researchers had asked that question. Ballard, B. M. (2021). COVID and CAFOs: How a federal livestock welfare statute may prevent the next pandemic. *North Carolina Law Review, 100*(1), 281.

43. Constance, D. H., & Bonanno, A. (1999). CAFO controversy in the Texas Panhandle region: The environmental crisis of hog production. *Culture & Agriculture, 21*(1), 14–26.

44. Head III, T. R. (1999). Local Regulation of Animal Feeding Operations: Concerns Limits, and Options for Southeastern States. *Environmental Law, 6*, 503.

45. Ogishi, A., Metcalfe, M. R., & Zilberman, D. (2002). Animal waste policy: reforms to improve environmental quality. *Choices, 17*(316-2016-7143).

46. Graham, J. P., & Nachman, K. E. (2010). Managing waste from confined animal feeding operations in the United States: the need for sanitary reform. *Journal of Water and Health, 8*(4), 646–670. Moses, A., & Tomaselli, P. (2017). Industrial animal agriculture in the United States: Concentrated animal feeding operations (CAFOs). In *International Farm Animal, Wildlife and Food Safety Law* (pp. 185–214). Springer,

Cham. Cerussi, A. (2020). Animals in agriculture: Federal legislation introduced to address problems with CAFOs. *Animal Law*, 9–11.

47. His name has been changed to protect his family's privacy.

Chapter 4

Filming in Galveston

Making a film is better than watching the film you just made. Watching your own film, especially with an audience, fosters a sense of anxiety, terror, and dread. They are going to hate it. They are going to think I am stupid. I will never get to make another film.[1] The salvo of thoughts is like bombs in your mind. It is hard to not think about what you could have done better. If only? If only more time, more money, better words . . . the soundtrack of life.

I had other projects to wrap up before I could fully devote myself to swine. The documentary film series on water was very important to me and I wanted to do a great job. I had written, produced, and directed this series. And, it has been distributed around the world. It remains the best-seller of all my documentary films.[2] But I still needed to finish some interviews, tighten up, the editing, and adjust all the b-roll (background footage). I had a couple of stops, and then I was heading to Galveston, Texas.

I needed to film in Galveston for many reasons. Galveston is an island in the Gulf of Mexico south of Houston. According to legend, the Spanish called it the "Isle of Doom."[3] It was once an economic powerhouse much bigger than Houston until the Hurricane of 1900 wiped homes and industries into the Gulf.[4] Yet the beauty of the city remains in homes and gardens that reflect an earlier, more prosperous time.[5] I was there to film water quality issues, such as industrial and municipal discharges, not to reminisce about the past.

Before I could get on the road again with the film crew, Juan and I spent a very hot summer at the Swine Unit. Summer was ideal because I would not be teaching my classes. I had a three-month "break" to solve an international water problem. CAFOs were impacting drinking water across the US, and that meant sluggish development.[6] Legally, states were grappling with how to administer federal law. Should this be a *Clean Water Act* (CWA) permitting issue regulated by the states? Or should this be a state process in collaboration with the USDA?[7] Sure, everyone would agree that swine CAFOs demanded reform, but what about aquatic farms?[8] The issues, like the murky waters of a dirty lagoon, seemed increasingly dark, and gloomy. And the politics around the issue stunk!

The task was herculean.

But we had a paying client that had invested his money and a basketball star's money in a project, and they wanted results. We knew their previous investment into a cyanide-based disinfectant had proven unwise. In fact, it was just a hustle by Richard and Guidry. We were diligently experimenting and building the system. Because of the limits placed on use of their space, LSU wanted us to have a small footprint and stay out of their way. Again, we leased a trailer. But the first real breakthrough came when we realized that to manage the waste, we had to intercept the waste before it went to the lagoon (see Figure 4.1).

The idea was really very simple. But, like all good ideas, they only seem simple in hindsight. More research has shown the importance of segregating the waste according to the age of the pig, which reflects the diet and thus the nutrient content of the waste stream.[9] Much research has been conducted on treating lagoons, regulating, lagoons, and reuse of the lagoon waters.[10] Recycling lagoon waters and reusing for washing down the pigs was a dream for many researchers.[11] This was theoretically possible once the solids sunk to the bottom and the top layer of water would be filtered and reused.

Figure 4.1. The Holding Tank, where all the waste was diverted from the Lagoon. Courtesy of the Author.

Capturing the energy value of the solids portion meant biomass to bio-gas.[12] Europeans, early adopters of digester technology (from China to the Netherlands), found this idea appealing as it would satisfy the water and solids management issues while generating electricity.[13] Yet, the tank was so big, we could not lift it. "Volunteers" from Dixon Correctional Facility (a nearby prison that did work for LSU) made that possible.

The holding tank solved many issues. The flow of waste to the lagoon was often too slow as the system relied on gravity and not mechanical pumps. Further, the Swine Unit rarely washed down the pigs; thus, the concentration of waste was higher. In other words, the hog waste was thick and slow. In large commercial operations, pig waste flowed, and the lagoon dimensions were massive. Some CAFOs used drying beds or even mesh wiring systems to reduce the moisture before further treatment.[14] Another option was dewatering the waste through a centrifuge system.[15] In either case, the water would require further treatment to eliminate the ammonia or other nutrients and metals before being land-applied or sprayed on the crops or open fields.[16] But we were not there yet. In fact, we were still a long ways away.

First, we never had more than 300 pigs. Your average CAFO numbered greater than 25,000 pigs. Our hogs were not farrowing, they were typically larger and older. Next, because there was little excess water, the solid waste portion was more like sludge. Of course, the urine mattered, but that was just another part of the sludge that needed to be treated. It did very little to dilute the poop.

Then the higher concentrations of ammonia (and other chemicals) meant we were not designing an alternative for the "real world" of pig crap. Also, the Swine Unit carefully monitored all diseases and that meant much higher antibiotics in the waste stream. This is an astonishing point. And not just for CAFOs. The presence of antibiotics in human sewers have been shown to slow down the bacteria needed in anerobic digestion during the phases of wastewater treatment.[17] The persistence of pharmaceuticals in wastewater is a significant concern for treatment plant operators.[18]

So, we needed to supplement the sludge with water. This could not come from a freshwater source, either. This had to be from the lagoon. Piping the water from the lagoon into the holding tank was not an issue. The issue was where would be the entry point. Would the lagoon water spray into the top? Or be measured into the bottom? How could we ensure a consistent volume of waste that was homogenous? We needed to build a system that was treating the same poop every day.

The optimal location was close to the center of the lagoon. Here, the water would have less solids and could be easier controlled. But it also meant extending the piping system towards the center (see Figure 4.2). We rented a small boat and hired an electrician to run the pumping to the middle of the

Figure 4.2. Lagoon Water Piping, where excess fluids were pumped into the Holding Tank. Courtesy of the Author.

lagoon (see Figure 4.3). Soon, we would have a steady flow of solids mixed with lagoon water that approximated the concentration of CAFO hog waste. And it would look and smell just like the real thing. So, now what?

During the later 1990s, ozone technology was all the rage. Ozone had been successful at reducing air emissions (and the odors that accompanied them) during the previous decade.[19] Again, Chinese researchers had demonstrated the power of an extra molecule of oxygen in eradicating ammonia in pig waste streams.[20] As European researchers followed Asian efforts, ozone was a wonderful supplement to other technologies.[21] Japanese scientists were confirming these findings.[22] There was an added benefit of reducing bacteria and diseases in the waste as well.[23]

So, what was the issue? If it worked so well . . .

Ozone worked well but was expensive.[24]

And agriculture is hypersensitive to cost. Ozone could add dollars on every pig going to market.[25] Ozone alone would not be the solution. But it did not stop us from buying a very expensive ozone system and watching it fail. Ozone is a powerful disinfectant and has many applications; proper, safe handling is critical.[26] Many times the system would back-up and ozone would leak and scorch our eyes, lips, and noses. Still, not all was a loss.

Something was beginning to emerge: solving this pig poop puzzle would require a hodge-podge of technologies. Each one would have to solve part of the problem, and all would have to be cost-beneficial.

Figure 4.3. Electrician in Boat, where the pumping system was installed. Courtesy of the Author.

The next phase of the project would be building the treatment system. We had the supply of waste, enough lagoon water, and we even built a device that floated on top of the surface of the tank water to ensure that pig waste did not settle to the bottom. We wanted to treat the real stuff. Not the watery part that separated, or supernatant.

Juan's expertise was in using anerobic reactors. Since the pig waste lacked any oxygen, employing anerobic tanks filled with some type of media for the bacteria to make their homes made sense. I really had little to offer here. I had little expertise to counter his training and knowledge. However, I was able to think through the process and come up with the idea of having a single conical shaped tank instead of 8 to 12 little cones in the bottom. First, it was just more efficient. Second, it was less expensive and easier to build.

We needed help, though. Our workers were traveling men that picked up work where they could get it. Really the only type of people that would ever do what we needed. Juan and I went through about ten workers, but Wayne and Don were the most consistent and the ones there from the very beginning. Wayne was younger, wild black hair, and always carried some type of weapon (i.e., knife usually). Don was older, much smarter than everyone, and could not follow directions if it meant saving his own life. He was about 60 years old and had been on the road longer than anyone I had ever met. Yet they helped us build what became known as "Jurassic Park" (see Figure 4.4).

Figure 4.4. "Jurassic Park," named due to its ever-expanding, seemingly endless system. Courtesy of the Author.

This system received waste from the tank and lagoon, and then was treated in a series on anerobic reactors (initially dosed with ozone, but discontinued), and then emptied into a final chamber for further treatment. After trying many other media for the internal system, we remembered the use of Zeolite in chicken research. There was considerable research here and it also aided in odor reduction.[27]

Zeolite was perfect. The denitrifying bacteria quickly formed around the jagged surface, which meant an increased surface area that allowed for better, faster treatment. What we had done was like taking a sewer treatment plant and reducing the footprint. Instead of acres of ponds and lagoons, we had three reactors on the top of a trailer. Being on a trailer also had another advantage: our system was portable, movable to a CAFO to treat the lagoon or pig units and then move on.

There is no use in pretending that everything went "according to Hoyle" from start to finish. It most certainly did not. Besides the regular pressures of life, family, work, and the incessant meddling by Jordan Hill (his right as the lead "investor"), the research was very taxing. Juan and I would argue and argue and argue. Never losing respect for one another. But quarreling in the heat covered in pig shit can try any professional relationship.

Troubled times and dark clouds were coming. And the foul sky does not clear without a storm.[28] We heard the thunder and could smell the effects of lightening. We thought these days were rugged. Those days at the Swine Unit would be nothing. Being covered in poop and showering outdoors before going home would seem like easy work compared to later.

It is amazing what you think you know.

Still, every day meant being soaked in feces, roach droppings, and sweat.[29] It was disgusting, and we earned every nickel. The setbacks would have derailed almost anyone else. We had a sort of messianic zeal to solving this environmental enigma. Our funders picked up on our excitement and that motivated them to anticipate every innovation and fall into trepidation with any setback.

There was, nonetheless, a conundrum. Again, there were two issues: water and solids. But, as we cleaned up the water, it produced more sludge, which meant more solids. If we cleaned up the solids, we had more water (because it required removing so much water). We chased our ecological tails for weeks. Often, we had strange results. Once it produced this dark foam that was high in nutrients. I used it around my home to "naturally" fertilize my flowers. The blooms were beautiful, colorful, and full. The smell was horrific. My neighbors assumed a sewer line had burst and called the Public Works Department to conduct a smoke test.

That was a failure.

Or was it?

We needed to change our thinking (again).

What if we dewatered the solids and made a fertilizer pellet? Or disinfected the solids and made a food pellet? What about biogas? Maybe the waste had BTU value. We through ourselves into solving the water part of the equation first (see Figure 4.5). Compared to the thick, black water we treated; this was a discovery that would change the CAFO industry.

I could now get on the road to Galveston and finish filming knowing that we had "solved" half of the problem. As we packed up the gear, I was already thinking ahead to the solids treatment. I understood compost management. I understood the reuse and recycling of solids in agricultural settings. I knew that we were on to something extraordinarily big. As I drove, I stayed on the cell phone with Juan talking about the next step.

As we wrapped up an exciting day of filming and were settling into the nightly pre-production work of watching the "dailies" (what we had filmed earlier), one of the locals asked if he could join us. He seemed pleasant enough. Sometimes you get strange requests when you show up with a camera crew. Some people imagine themselves as part of the crew and start offering direction and production ideas. Others just want to see how the "magic" happens. This guy was one of the latter. I think his name was Bryon.

"I bet you never get bored," Byron seemed to be asking.

"You would be shocked," I yawned.

Figure 4.5. The Clean Water, the results of so much effort to produce potable water. Courtesy of the Author.

"What other films have you done?"

"Garbage, recycling, incinerators, compost, and lots of sewer and industrial plants. Pretty much anything environmental," I remarked.

Silence.

We watched more dailies. Usually, this part of the day was more relaxed unless the interviews were awful, or the b-roll was dreadful. These times called for a stiff drink. Beer was never enough. These were Scotch or Tequila times.

Our luck had been good. These shots were very good. We had a great one of a deer playing next to a formerly polluted pond. Another showed a bird swooping down and grabbing a fish while we were interviewing an activist about the water quality.

You live for shots like that. As a note, and for full transparency, we drank more when the shots were good than bad. We ate much more as well.

"Man" he added, "You really have an interesting life!"

"It only seems that way," I remarked.

"You do work in all these areas."

I nodded. I hoped it was interesting and meaningful.

He then said something that changed my life forever.

At first, it was understated. Like some many irrelevant comments you hear at another worthless party. Mundane, boring comments. I wanted to make sure I understood him.

"Do you ever do research with pig waste? You know, like CAFOs?"

The question hung in the air.

"Why?"

I was astonished. His question had hit the mark. What should I say?

He continued, "I have been working with this group of scientists, farmers, and investors from Belgium.[30] They are looking for alternatives. Can I get them to give you a call next week? They are serious people. They are real."

Is this how fortune finds you?

Is this how success makes its way to your door?

Serendipity.

"Have them call me," almost grudgingly with a hint of sarcasm and disbelief.

I gave Byron my home telephone number and asked him to pass it along to the group. I knew that I would be at work whenever they called so I had an excuse to not be available. I figured my wife would filter out any kooks.

My life was about to change.

NOTES

1. I have always felt terribly nervous when showing my films. It seems to get worse and worse.

2. Films Media Group. Our urban environment: Water quality. Located at https://www.films.com/ecTitleDetail.aspx?TitleID=4969. Accessed 20 April 2022. This was my second film series. The eight-part series on garbage had come out the previous year. Ultimately, I produced, directed, and wrote four film series on several environmental subjects. One of the best dealt with ecological challenges for shipyards. It was the coolest, most high-tech film. It never saw the inside of a screening room. The funders snuffed it out before it could be shown to a mass audience.

3. McComb, D. G. (1986). *Galveston: A history*. University of Texas Press. This is one of the more definitive books on this fascinating town.

4. Cartwright, G. (1998). *Galveston: A history of the island* (No. 18). TCU Press.

5. Gray, E. B. (2016). *Greetings from Galveston: A history from the 1870s to the 1950s through post cards*. Lulu Press, Inc.

6. Jones, R., Frarey, L., Bouzaher, A., Johnson, S., & Neibergs, S. (1993). *Livestock and the environment: A national pilot project. Detailed problem statement*. Tarleton State University and Iowa State University.

7. Fare, G., & Potts, S. E. (1987). Rising federal waters: The nation's quagmire. *Baylor Texas Environmental Law Journal, 18*, 1. This was one of the first law school articles to take on this subject.

8. Barker, A. P., & Burleigh, R. B. (1993). Agricultural chemicals and groundwater protection: Navigating the complex web of regulatory controls. *Idaho L. Rev., 30,* 443. Agriculture has long been able to skirt environmental regulations. It is simply hard to enforce laws and guarantee compliance in the country.

9. Brooks, J. P., Adeli, A., & McLaughlin, M. R. (2014). Microbial ecology, bacterial pathogens, and antibiotic resistant genes in swine manure wastewater as influenced by three swine management systems. *Water Research, 57*, 96–103.

10. Jones et al.

11. Westerman, P. W., Safley, L. M., & Barker, J. C. (1990). Lagoon liquid nutrient variation over four years for lagoons with recycle systems. In *Agricultural and Food Processing Waste: Proceedings of the 6th International Symposium on Agricultural and Food Processing Wastes*, December 17–18, 1990, Chicago, USA (pp. 41–49). American Society of Agricultural Engineers. This became a guide throughout the project. So much information can be gleaned from assemblies of experts. Often, conferences degrade into a "spring break" for academicians and researchers.

12. Taiganides, E. P. (1982). *Biomass from the treatment of pig wastes.* Wiss. Umwelt ISU; (Germany, Federal Republic of), 4.

13. Safley Jr, L. M., & Westerman, P. W. (1988). Biogas production from anaerobic lagoons. *Biological Wastes, 23*(3), 181–193. This article steered us towards looking for ways to capture the methane produced when the solids fraction decomposed.

14. Choo, P. Y., Teoh, S. S., Lim, Y. S., & Ong, H. K. (1989). Drying of pig waste lagoon sludge on mesh-lined timber beds. In *International symposium on waste management and recycling in pig farms*, Singapore, 9–10 Oct 1989. [Primary Production Dept.].

15. Miner, J. R., Goh, A. C., & Taiganides, E. P. (1983). Dewatering anaerobic swine manure lagoon sludge using a decanter centrifuge. *Transactions of the ASAE, 26*(5), 1486–1489.

16. Paul Taiganides, E. (1986). Animal farming effluent problems—an integrated approach: Resource recovery in large scale pig farming. *Water Science and Technology, 18*(3), 47–55.

17. Gallert, C., Fund, K., & Winter, J. (2005). Antibiotic resistance of bacteria in raw and biologically treated sewage and in groundwater below leaking sewers. *Applied Microbiology and Biotechnology, 69*(1), 106–112.

18. Polesel, F., Andersen, H. R., Trapp, S., & Plósz, B. G. (2016). Removal of antibiotics in biological wastewater treatment systems: A critical assessment using the activated sludge modeling framework for xenobiotics (ASM-X). *Environmental Science & Technology, 50*(19), 10316–10334.

19. Wong, K. W. (1986). Use of ozone in the treatment of water for potable purposes. *Water Science and Technology, 18*(3), 95.

20. Zhang, R., Yamamoto, T., & Bundy, D. S. (1996). Control of ammonia and odors in animal houses by a ferroelectric plasma reactor. *IEEE Transactions on Industry Applications, 32*(1), 113–117.

21. Hobbs, P. J., Misselbrook, T. H., & Cumby, T. R. (1999). Production and emission of odours and gases from ageing pig waste. *Journal of Agricultural Engineering Research, 72*(3), 291–298.

22. Inaba, M. (1991). Utility of ozone for waste management in pig farming. Bulletin of Shizuoka Swine and Poultry Experiment Station (Japan).

23. Turner, C., & Burton, C. H. (1997). The inactivation of viruses in pig slurries: A review. *Bioresource Technology, 61*(1), 9–20.

24. Wu, J. J. J. (1998). *The use of ozone for the removal of odor from swine manure.* Michigan State University.

25. Adams, R. M., Glyer, J. D., Johnson, S. L., & McCarl, B. A. (1989). A reassessment of the economic effects of ozone on US agriculture. *Japca, 39*(7), 960–968.

26. Moore, G., Griffith, C., & Peters, A. (2000). Bactericidal properties of ozone and its potential application as a terminal disinfectant. *Journal of Food Protection, 63*(8), 1100–1106.

27. Tavolaro, A., & Drioli, E. (1999). Zeolite membranes. *Advanced Materials, 11*(12), 975–996.

28. All apologies to Shakespeare. *King John.* Act 4, Scene 2.

29. Once when cleaning a pipe, the line broke and Juan and I were covered in roaches. Thousands of them. I am very afraid of roaches, and this was a real horror show for me.

30. Belgium sponsored lots of research and collaborated with Germany, France, and the Netherlands. Demuynck, M., Honnay, J. P., Neukermans, G., Bels, J., & Assche, P. V. (1984). Methane production from pig and cattle slurry in Belgium. *Revue de l'Agriculture, 37*(2), 197–210. Many wanted to return to family farms, especially along the Flemish coast of Belgium. Maton, A. (1990). Economic, management and technical aspects of pig keeping on a Belgian family farm. *Revue de l'Agriculture, 43*(5), 809–826. That era was long over. Instead, the focus was shifting towards larger American-like CAFOs and how to manage the waste after it was generated and less about waste reduction. Poels, J., Verstraete, W., Neukermans, G., & Debruyckere, M. (1984). Biogas produced from the liquid manure from pigs-1st practical results from a large-scale installation. *Review of Agriculture* (Brussels);(Belgium), *37*(1).

Chapter 5

The Man from Bruges

There are hustlers everywhere. The environmental sector seems to bring out the most insane con-artists. The attraction to turning "good into gold" seems too alluring for most to resist.[1] I have seen so many looney ideas for saving the planet that I am almost immune. But that is today. Back then, I was all enthusiasm and way too imprudent. Maybe I was just naïve.[2] It is difficult to remember. Further, even a great idea takes much effort to get an industry to adopt it.[3] For agriculture, especially in price-sensitive pig farming, there were tons of hurdles to prevent, delay, or avoid technological innovations.[4] I do not think we ever realized how far away we ever were.

While living in Baton Rouge, so I could be closer to the Swine Unit at the Ben Hur Research Station on the LSU campus, I met several "entrepreneurs" that wanted to "get in on the action" on this "whole pig thing" that I was working on. I really had no idea that such people existed. Patent and research bottom-feeders looking to profit from brighter people's hard work and brains. Universities are often the target and the perpetrators of patent trolling.[5] Some people have the money, but no ability to create or invent something special. To be fair, I was not so sure that I could make this happen either.

The discussions were direct and candid.

"What will it cost us to buy a portion of your future proceeds?" One of them asked.

"I don't know," I said, "I have really never thought about it."

And I had not thought anything about it. I was too busy trying to invest something and solve a problem. Of course, I had an idea of what a patent could be worth.

"We have a group of us that would like to offer you a chance to get some of your reward now for your hard work and brains." They pleaded.

They were right.

The project was paying bills. Partially. My job at Tulane University would start-up again in a few weeks. I was not hurting. But I had a small child, a wife that had dedicated herself to raising our child and left-over bills from my mother's cancer treatments and extended hospital stays. Any additional money would be welcomed. I guess they knew that. More than likely, they were counting on that.

I asked, "What do you value the pig invention to be worth?"

I really had no idea at this point. I knew what intellectual property meant. But we did not have anything like that, yet. Following the success of the water treatment, solids management was now insight. We met at a cigar bar in Baton Rouge. Plush leather chairs, fine whiskeys, bourbons, and lots of Scotch. Laughing, smiling businessmen and politicians. Tons of backslapping and dirty jokes. Is that not the proper environment for deal-making? I took the napkin under my glass of Scotch, a 12-year-old Balvenie, took a pen from my pocket, and drew a series of tanks and pipes. Waste comes in and potable water goes out.

Very basic picture with my limited artistic abilities.

I offered, "This is what I am thinking and that is as far as I have gotten so far."

They must have been impressed. They were even more impressed with the fact that a couple of weeks earlier, one of the largest agricultural chemical companies in the world flew to Baton Rouge to "check out the invention." The reason this group of investors knew about this is because I had asked one of them to help me paint the tanks black (see Figure 5.1).

Figure 5.1. The Tank Close-Up, a candid picture of the treatment tanks. Courtesy of the Author.

The reason for painting the tanks black was obvious. One, I figured this would increase the internal temperature of the tanks and make the bacteria work faster. It did. The tanks soaked up the sun and the internal temperature was much higher. Two, I did not want this large company to see the internal workings of the system. The tanks were impossible to see into with the black paint.

I did not want this company stealing our technology before we could patent it or protect it in some other way. Of course, I did not trust them. As far as I was concerned, they had the money, the technology, and the political connections to fix this problem. They did little. Not everyone wants a problem solved. Some people make money perpetuating the problem. Maybe that is where they were?

I will never forget the response from their lead scientist.

"So, we spent $25 million trying to solve this problem and you boys did all this for under $300,000," he asserted with skepticism and sarcasm.

I was already sick of him. "Sir, do you really want me to comment on the incompetency of you and your staff?" I demanded.

They went back to Missouri empty handed. None of us liked any of them. Arrogant and ignorant. Two dangerous combinations.

Back at the cigar bar, I was meeting with this group of investors.

They proposed something strange. They had probably discussed all of this before the ice cracked in the first glass of Scotch.

"Let's just say the entire value is $15 million. You have Jordan Hill, and Juan Josse. That means your part could be worth $5 million."

"Ok."

I was listening. I did not drink another drop. They were talking real money. I was not going to make a deal while intoxicated.

"Then, if your part is worth $5 million, and you assigned us, say, 3 or 5 percent of you portion, then that could have a value. There is still much uncertainty."

"Maybe more than you realize."

I was trying to be honest and open. We were still a long ways away from having anything of significant value.

"Sure, but if we paid you $100,000 for 5 percent . . . would that at least be a place to start?"

"I need to think about this."

"Do so. Then, call us back."

We clinked glasses, cut another cigar, and talked about other things. My mind stayed fixated on the moment.

Was this real?

Just a few weeks earlier, an offensive and insulting offer from another group had been made. In that offer, it was proposed that Juan and I would give up majority control of the project for no money but the promise to build a full-scale unit at some point in the future. That group had approached me through a family member. It was disgusting from the first meeting to the eventual "Hell no!" I have found that people use all sorts of tricks to lure you into a meeting and then pressure you to make a deal that favors them and not you. These people must have assumed that Juan and I were more than just gullible and naïve.

The group in Baton Rouge was hitting closer to home. At least I was not insulted. And the scenery was better. We had met at a cigar bar in Baton Rouge. Not in an old warehouse full of machines and chemicals. When I got home, I reeked of smoke. Considering that I had ruined many dinners stinking like pig waste, this was a welcomed relief.

"How did it go?" My wife inquired.

She was naturally skeptical because one of the people I met with that night was a former college roommate with little to show for his efforts including seldom having a stable source of income or even a job. He was probably waiting for his rich dad to die. This is the flaw of generation inheritance.

"He hasn't changed, but the group was interesting," I responded.

"How so?"

"They want to buy into the pig deal."

"With what?"

"The group has money. Or so they say. It did get me to start thinking about what comes next. One of them has an uncle that is looking to invest in something for the long-term. A couple of the guys have a business and need a place to put some of their profits. I don't know yet, but it got me to thinking."

We talked a little more, I showered, and turned in for the night. Washing off cigar smoke was easier than scrubbing off skin and hair trying to erase the daily filth from the hog farm.

The next day, I had to work in New Orleans. As expected, the man from Bruges called.

My wife spoke briefly to him. Then, she called me at my office.

"There is a dude from Belgium that called. He said a man from Galveston, Texas named Byron told him to call you. His accent is strong, but he sounds legitimate."

He cleared the first hurdle.

My wife was my filter for wackos like this. She still serves this role. Because she is naturally skeptical and worked at the Louisiana Department of Environmental Quality (DEQ), she had the expertise and experience to evaluate "proposals." And more patience than me. If she thought he was worth a second look, that was all I needed. To be honest, I was in shock. Lots of people had an interest in this issue. I think many people were curious. Like watching an accident on the side of the road. Intrigued, but were not about to stop and do something about it. This was something more than that.

Could a random question in Galveston lead to triumph?

Was this real?

I immediately went to Juan's office, and we discussed the matter.

"John," he said seriously, "We are heading to North Carolina soon. We still have much to do."

I agreed, of course.

But the time for thinking ahead was now.

"How about if I speak to him and check him out?" I requested.

Juan agreed. I generally read people well and could sort through their hog wash faster than him.

So, I did.

Around 2 am my time, but Belgian daytime, Reginald de Vos (called himself D. Reggie) telephoned. He was a Belgian national with international business connections. What that meant was far from what we thought. D. Reggie loved good food, expensive wine, and fast European cars. So far, so good. But he had not worked a regular job in more than a decade. He had some interests that took him to Eastern and Central Europe, and he has an "agent" for large multinational corporations (MNCs).[6] He claimed to have once served in the Belgian army and was a technology and logistics expert. He mumbled as he spoke and never seemed totally honest. Over the telephone, though, I could not pick up any of these cues.

Later Juan and I realized that he was a version of Jordan Hill but with a Flemish accent. His advantage was that he read and was curious. He could understand some of the research. He had put together enough deals to warrant a closer look. Also, D. Reggie had worked with British agriculture experts and knew people across the Chanell with connections to large farming operations.[7] Additionally, Reg knew that environmental considerations in Europe would be the dominate matrix for determining success and adoption of any new technology.[8] Unlike the US, Europeans emphasized ecology over economics.[9] The model for success would be different on that side of the Atlantic.[10] In the early 1990s, the European Union (EU)[11] had pushed the notion of sustainable development into almost every industrial, commercial, and, now, agricultural sector.[12] This concept is simple enough: businesses and consumers should consider the long-term impacts of their actions and not just immediate profits.

"I am looking to invest in technology and environmental solutions that solve this pig problem," he explained.

He was part of a consortium of financial and agricultural people that had interests throughout Europe. D. Reggie knew people. He had friends.

"How soon can you and your partners get here?"

I explained that we were tied to Jordan Hill (at least at that moment) and had a commitment to go to North Carolina.[13] North Carolina always seemed a better option than Iowa because of so many water-related pollution and contamination concerns and their environmental justice problems that were connected.[14] Besides, we wanted to check out an American CAFO first. If the technology was cost-effective in the US and environmentally sustainable according to European standards, we really did have something.[15] All boxes could be checked, and the project could be deemed viable. Then, going to Belgium or any other country in the EU would be much easier. And D. Reggie wanted to review some options closer to home, such as diet management and odor reduction.[16]

That seemed fine. We knew that there was no long-term solution to be found in chasing food and diet for pigs. The problem remained: too many pigs in too small of an area producing too much waste for the local water and soil ecologies.

On this side of the Atlantic, Jordan Hill was getting frustrated and a bit uneasy. He had run out of time, money, and could not find either one anywhere else. He did not want to lose control of the project by having us go to Belgium with financial interests and connections.

He pleaded, "Let's discuss all this when you get back from North Carolina."

"Do you have a farm ready for us?" I asked.

Silence.

Nothing.

Jordan Hill was supposed to arrange tours of CAFOs in North Carolina. This is what he was supposed to do between cigarettes, coffee, and religious TV shows from his comfy leather chair in his air-conditioned living room. He had visited the Swine Unit only once. He did not like the smell. It was now clear. He had done nothing more than make some initial telephone calls, make spaghetti, and offer excuses.

We were in a jam now. Well, who would set up the tours?

He was supposed to set up a meeting with people ready to adopt our emerging technology. He did neither. He did not have a single lead or a single farmer that spoke with him more than a few minutes. I was afraid his insanity and prejudice were coming through the telephone lines. I needed to take control of this.

So, I started "cold-calling" pig farmers until I found a group of brothers with about 500,000 pigs. They had divided them into three separate farms to escape having to comply with North Carolina's generous CAFO regulations. They knew they had a problem due to the proximity of a river and complaints of neighbors.

They could see the future.

The future would have more regulations and actual enforcement.

They had already tried the enzyme scam, the air scrubbing scam, and a few others that we had not heard of yet.[17] I had seen powders, worms, and liquids and tons of other products poured into the lagoons that offered nothing. Many of these types of products are still around.[18]

Juan and I travelled to North Carolina and inspected the facilities.

They were ideal.

Lots of pigs. Lots of waste. Pissed-off neighbors.

We called Jordan Hill.

He was not excited.

He would not commit.

Something was going terribly wrong here. Why did he not want to follow through with this? His part was merely setting up an appointment. Juan and I had to go "sell" the idea.

What was the purpose of all the research if not to put into practice in an actual farm? Jordan was broke. He could not honor his commitments. His commitments were subject to a successful demonstration, based on independent laboratory analysis of the water (not the solids), he would invest more into a full-scale model. He simply could not do it. His dreams were too big for his reality.

It is a real shame, too. Despite his crazy racist, homophobic nonsense, he had jump-started this project with an initial investment. I felt bad for him. But not bad enough to let him kill our efforts. He wanted to engage a lawyer from St. Louis to help us sort through all the issues. Jordan made so many religious bigoted comments about the lawyer while in his presence that I could not believe he would still answer his calls.

The CAFO we toured was ideal. Call it luck. We visited one major farm and they were ready to sign a deal right then. Being hog farmers, they were naturally leery of university types like us. But we understood their concerns and found a way to make the numbers work. This could be a cost-beneficial solution where they could have access to the technology and set the industry standard. In short, they would have a competitive edge on their competitors.

Juan and I had a pleasant meal and were prepared to go back to New Orleans. Jordan Hill would have the opportunity to fulfil his side of the bargain: raise the money for a full-scale system at a CAFO in North Carolina.

Tragedy struck.

The day before Juan and I were heading back to Louisiana, I got sick.

Real sick.

The pain in my lower, left gut was excruciating. I could not eat without throwing up. I assumed that I had food poison. I sweated buckets and ran a high fever. I could barely walk. It hit me all at once. As we got on the plane, I decided to ask for a Coke and some peanuts. I was shaky from being so hungry and nauseated. I figured this would calm my stomach. I also requested an icepack. The flight attendants allowed me to lay down in one of the bathrooms with my feet hanging in the toilet. Even when we landed, I stayed on the floor. They must have seen the death grip that I was in. Or maybe I still smelled like the pig CAFOs that I had visited.

When we landed in New Orleans, I made my way to the baggage claim bent over at the waist. I could not straighten up without experiencing a pounding in my stomach. My forehead was drenched in sweat, and I was dizzy. It was like eating County Fair corndogs and then being on a cheap rollercoaster.

"What did you do in North Carolina?" my wife inquired. No doubt she suspected that Juan and I had enjoyed too much of the Tar Heel state.

This was not from drinking too much or too much food.

I asked that I be allowed to rest for a few hours before taking me to the emergency room. I needed to try to rest for at least a couple of hours first. I had not slept for more than a day and I needed my strength.

By 4 am, I was ready to go. I could hardly see, and my fever had passed the century mark. I was getting delirious.

At some point, the doctors diagnosed me with acute diverticulitis.[19] This is what had been linked to the death of both of my grandfathers. It was nothing to play with. The worst thing I could have done was flying in a plane, drinking a Coke, and eating peanuts. I was lucky to be alive.

I would be hospitalized and unable to go to with Juan to meet D. Reggie in Belgium.[20] I had more pressing matters. For the next 30 days, I survived on a fiber-free diet, lots of water, no alcohol, no tobacco, no coffee, and the knowledge that a major surgery was waiting for me.

Juan would have to represent us in Belgium.

He visited he before left and immediately after he got back.

"John," in his serious tone, "These people are real, and we should work with them. I would even move there if needed. They want us to come back in a few weeks."

"What about Jordan Hill?" I asked, both of us knew the answer already.

We began planning to go to Europe within a few weeks after my surgery.

I spoke to D. Reggie a few times over the next several weeks. Fortunately, I spoke and met with Jordan Hill less and less. Ultimately, he must have realized that his participation in this project was finished. Not even his "smart-duck Jewish lawyer" (as he called him) could not save him. Hill was out of the hog waste business. So, he started a "Christian clothing store" and a few more companies. Within a few months, any pretense of a relationship was over.

My brush with death changed me.

I knew that my time was limited.

There were no guarantees.

I never felt the need to suffer fools again.

Life is too short.

I learned a valuable lesson: ask people what the budget is for a project. If they do not have one or have no money, then it sounds too much like coffee shop conversations that lead to more coffee drinking and rarely to anything substantive. And if I must come up with money for someone's idea, then it is probably not much of an idea. It takes more than a good idea for something to become real.

A sharp dealmaker once told me, "If a man will not mortgage his home for his idea, then he does not really believe in it himself. He wants you to subsidize his risk-taking."

How true.

Also, more than anything else, I started thinking about what I would leave my family. I had bought insurance, but I wanted them to have something more.

I worked out a deal with the cigar bar investor group. To my surprise, they secured enough money to justify a negotiated deal. They even made money themselves by selling their "shares" to others.

Now I had some pocket money, my health was returning, and I was ready for Europe. D. Reggie here I come!

Juan and I would be partners with international businessmen with connections.

The money would pour in!

So, I thought.

Trading a broke American for a group of crooked Europeans was not a step in the right direction.

Oh, what I did not know!

I was going to get a lesson in international dealmaking.[21]

What is the quote from Proverbs, "Pride cometh before the fall?"[22]

Arrogance? Probably.

Upon conducting a recent patent search for the period between 1980 and 1999, more than 15,000 patents were filed claiming to solve the problem.[23] More would be coming. We needed to make a deal and get started. Going from a trailer mounted "Jurassic Park" to a full-scale system would take lots of money and many more technical issues to resolve. Economies of scale do not always work in your favor when going from laboratory to demonstration to system.

We needed to act fast.

The clock was ticking.

NOTES

1. Gorman, M. E., & Mehalik, M. M. (2002). Turning good into gold: A comparative study of two environmental invention networks. *Science, Technology, & Human Values, 27*(4), 499–529.

2. Konnikova, M. (2017). *The confidence game: Why we fall for it . . . Every time*. Penguin. People seem to always fall for con artists. Otherwise, honest men and women would rule the world. They don't. I was looking for angels in a world populated with demons.

3. Barbieri, N., Ghisetti, C., Gilli, M., Marin, G., & Nicolli, F. (2016). A survey of the literature on environmental innovation based on main path analysis. *Journal of Economic Surveys, 30*(3), 596–623.

4. Possas, M. L., Salles-Filho, S., & da Silveira, J. (1996). An evolutionary approach to technological innovation in agriculture: some preliminary remarks. *Research Policy, 25*(6), 933–945.

5. Lemley, M. A. (2007). Are universities patent trolls. *Fordham Intellectual Property Media & Entertainment Law Journal, 18*, 611.

6. Stopford, J. (1998). Multinational corporations. *Foreign Policy*, 12–24.

7. Pain, B. F., & Klarenbeek, J. V. (1988). Anglo-Dutch experiments on odour and ammonia emissions from landspreading livestock wastes (No. 88–2). IMAG.

8. Jongbloed, A. W., & Henkens, C. H. (1996). *Environmental concerns of using animal manure—the Dutch case. Nutrient Management of Food Animals to Enhance and Protect Environment*. Lewis Publishers, 315–332.

9. Ibid.

10. Donham, K. J. (1998). *The impact of industrial swine production. Pigs, profits, and rural communities*, 73.

11. Liberatore, A. (1997). *The integration of sustainable development objectives into EU policymaking* (pp. 107–126). London.

12. Mitlin, D. (1992). Sustainable development: A guide to the literature. *Environment and Urbanization, 4*(1), 111–124.

13. Furuseth, O. J. (1997). Restructuring of hog farming in North Carolina: explosion and implosion. *The Professional Geographer, 49*(4), 391–403.

14. Davies, P. R., Morrow, W. M., Jones, F. T., Deen, J., Fedorka-Cray, P. J., & Harris, I. T. (1997). Prevalence of Salmonella in finishing swine raised in different production systems in North Carolina, USA. *Epidemiology & Infection, 119*(2), 237–244. To be fair, Texas and Oklahoma were becoming a choice destination for CAFOs, but still nothing like North Carolina. Frarey, L. C., & Pratt, S. J. (1995). Environmental regulation of livestock production operations. *Natural Resources & Environment, 9*(3), 8–12. But there was another reason for North Carolina. The state had been the focus of environmental justice/racism for decades. Working to solve these issues coupled with the prospects of purifying pig poop was appealing. Sutherland, E. J. (1999). The siting of concentrated animal feeding operations (CAFOs): Information gaps for achieving environmental justice. Georgia Institute of Technology.

15. Blount, G. W., Henderson, D. A., & Cline, D. S. (1999). New Nonpoint Source Battleground: Concentrated Animal Feeding Operations (continued), The Natural Resources and Environment., 14, 68.

16. Kornegay, E. T., Harper, A. F., Jones, R. D., & Boyd, L. J. (1997). Environmental nutrition: Nutrient management strategies to reduce nutrient excretion of swine. *The Professional Animal Scientist, 13*(3), 99–111.

17. Bonmati, A., Flotats, X., Mateu, L., & Campos, E. (2001). Study of thermal hydrolysis as a pretreatment to mesophilic anaerobic digestion of pig slurry. *Water Science and Technology, 44*(4), 109–116.

18. Aira, M., Monroy, F., & Domínguez, J. (2007). Earthworms strongly modify microbial biomass and activity triggering enzymatic activities during vermicomposting independently of the application rates of pig slurry. *Science of the Total Environment, 385*(1–3), 252–261.

19. Ferzoco, L. B., Raptopoulos, V., & Silen, W. (1998). Acute diverticulitis. *New England Journal of Medicine, 338*(21), 1521–1526.

20. The process of the diagnosis was awful. I am allergic to Barium. That is what you drink before you get an MRI so the doctors can check out your insides. I had never had an MRI or Barium before then. My breathing labored and I got very dizzy. I passed out inside the unit. I woke up in pain and needing to throw up again. Since it was colon related, I could not have any pain medicine either. Post-surgery was even worse. I felt like I was on fire. Then, shortly after the surgery, I got salmonella poisoning. My stitches ripped open, and I get my gut in with a small throw-pillow.

21. I needed a lesson here. Greider, W. (1997). *One world, ready or not: The manic logic of global capitalism*. Simon and Schuster. I should have read this before getting on the plane and heading to Brussels. Another wonderful work. Although it focuses on a certain ethnic group, the applications and lessons are too close to home. Marlock, D., & Dowling, J. (1994). *License to steal: Traveling con artists: Their games, their rules—your money*. Paladin Press.

22. Proverbs 16:18.

23. A Google Scholar search conducted on 18 June 2022 revealed more than 15,000 patents. Probably many of these were cross over from other agricultural waste streams and some probably duplications. So, let's just say that there were only 10,000 actual patents. The odds were totally against us. To add some perspective: you have better odds at rolling a seven in craps five times in a row than being the 1 in 10,000.

Chapter 6

The Boys from Europe

Too many Americans are so enamored with Europe that they forget why their ancestors left there in the first place. I was no better. The notion of doing business in London, Paris, and Brussels was alluring. The appeal of fast-paced luncheons in the City of Lights and hopping a train to Brussels before crossing the Channel on my way back to London was glamorous and captivating. I had travelled to Europe many times by then. But not in the circles I was about to step into.

Most people have been asked to respond to the 419 Nigerian Scam.[1] These are typical "Spanish Prisoner" cons where, for example, some innocent person has millions of dollars, sends a letter (or an email; nowadays a text), and promises to "release" the funds subject to a processing fee, a destination charge, or some other fraudulent trick.[2] This plays upon people's willingness to help someone in need (sympathy) and their desire to get something for nothing (greed).

W. C. Fields once quipped, "You can't cheat an honest man."[3] People still try.

Juan and I landed at the Luchthaven Zaventem (Brussels International) Airport in the fall of 2000. He had met "the Boys from Europe" a few weeks earlier and was impressed enough to warrant a return trip. I had not been able to come for the first visit as I was fighting the Angel of Death who stuck his wicked sickle through my guts and almost killed me.

Well, that is what the Angel of Death does. He has a job to do as well. I never held a grudge.

D. Reggie met us along with an elderly "gentleman" nicknamed "Shaky Eddie." Shaky Eddie had earned that moniker through decades of drinking wine for breakfast, beer for lunch, and gin for dinner. By 8 am, he was smoking like a two-stroke engine with too much oil to gas and his scratchy voice was like a muffler with rust holes. He was harmless, but he had taken a vow of no deodorant and he smelled worse that the focus on our research. D. Reggie smoked constantly and mumbled even more.

"I can afford to smoke," he stated proudly, "Nothing but these red label American cigarettes."[4]

"I thought Europeans taxed the hell of every vice. How can you afford to smoke so much?" I inquired.

"Ha! These are Chinese fakes with a US federal stamp. Just like the real thing. I don't pay any taxes"

"What?"

"Yep. I got a warehouse full of these. And, Tennessee whisky with an American government seal."[5]

"You are quite the entrepreneur," Juan added sarcastically.

The sarcasm was lost on D. Reggie.

"Got some cell phones from Finland, too. Whatever you guys want. I work with people at NATO, and I got all kinds of connections."

He mentioned a Japanese car dealership where we could have our pick while in the country. These were not fake cars. We could simply "borrow" them as needed.

Neither of us were impressed.

Across the table, Shaky Eddie finished off his "coffee" and lit another smoke. He could take a deep drag and leave almost a half an inch of ash hanging from the lit end of the stick.

We checked into a small hotel in the coastal town of Ostende. This is a sea-side Belgian town with great food, bars, and a wonderful view of the water.[6] Like many locally owned European hotels, this one was small, few rooms, and a wonderful restaurant. The staircase was extremely narrow. Juan and I took turns retrieving our luggage and checking into our rooms. The owner remarked that he did not get many Americans. Since Juan was from South America, neither of us offered any "technical" corrections.

"I used to get this black fellow years ago. Very quiet man. He loved to fish. Back in the early 1980s."

"That sounds good," I responded about as bored as possible. I was tired from travel, and a bit aggravated with Reg and Shaky Eddie. When you do not smoke and someone else seems hell-bent on blowing smoke in your face . . .

"The guy sang, too. I have a picture of us together. I would sneak off and we'd go fishing off the pier."

The stocky man lumbered back wiping his forehead and holding a faded picture. It was him many years earlier. The man with him was Marvin Gaye.[7] They had been out all day and had caught a couple of good-sized fish and were heading in to clean them before a celebratory meal. Apparently, Marvin Gaye had lived in Ostende in 1981. He needed to get off drugs and find himself and his music.[8] Marvin wrote the song *Sexual Healing* there. He and our innkeeper were fishing buddies.

I felt that fate had brought us to this place. Marvin Gaye's *What's Going On?* (1971) album was one of the reasons that I became an environmentalist. Both Juan and I were moved by this experience and talked about it many times.

The next day, D. Reggie, Shaky Eddie, Juan, and I toured the countryside and visited some pig farms. European CAFOs had less pigs and were more stringently regulated.[9] This was true for hogs and chickens.[10] Europe is just smaller, more densely populated and has stronger environmental regulations. The health concerns at the farm were amazing and we all had to wear protective gear to avoid spreading any diseases.[11] Shaky Eddie had to shower before being allowed near the hogs. He still stunk. I would swear under oath, but I think the pigs were offended by Shaky Eddie.

The facility certainly would be glad to allow us to build a full-scale system and get rid of their problem. Again, we had devised a method to clean-up the water (mainly urine and washdown water), but the solids fraction (the feces) was still up in the air like the odor issue. This had possibilities.

"Well, what do you think?" asked D. Reggie.

It would work out fine. But there was a catch.

"The farm does not want us doing any research near the pigs. We will have to ship the slurry, the waste to a processing plant."

Juan and I were stunned. That would create logistical and handling troubles. And an additional transportation expense. That meant liability insurance. Fuel costs. More labor. I was nervous. My anxiousness must have been unmistakable.

"Don't worry. I have a friend in England that knows logistics. He is the one that helps me get cigarettes and whisky into Spain and Portugal."

Wonderful. How comforting.

"D. Reg," I demanded, "We are going to need to do everything legally knowing that the government will be watching us every step. This must be done right. Whatever we do here must be able to be demonstrated to any official. This is not just a hustle."

He was not convinced.

We would have to persuade him.

After a shower and a couple hours of rest, we met for beers and smoked chicken at a nearby bar. Germans deservedly get credit for excellent beer. But Belgian beer is tasty and fresh. Flavored beers with smoked chicken and a cheese plate hit the spot. Sitting outside, listening to the waves of the sea, and talking into the night, I felt better about this project. We shared stories about our lives. My country life from Arkansas was foreign to them. D. Reggie brought an ex-pilot. Very bizarre night. Hard to remember everything that was said. Maybe it is time and the lapses of memory that come with age. Or maybe it was the beer.

We were scheduled to meet the rest of the "boys from Europe" within a couple of days. However, we detoured back to Brussels. This time, to brief the European Union (EU) and for their experts to evaluate our technology. Under the auspices of the EU are many areas of competency. One of those is the European Commission on Food, Farming, and Fisheries.[12] Any comparative analysis between EU and US farming practices demonstrated the level of competency and commitment the Europeans wholeheartedly attached to sustainable agriculture.[13] I remember feeling astounded and awestruck upon entering the EU. It is one thing to read and study about something. It is another to now be at the table with policymakers. Up the street from the EU is the Museum of Natural Sciences and a little coffee shop that had these wonderful little chocolates that were shaped like seashells. The meetings were informative, illustrative, and, sometimes, contentious. I should have known that one of the last vestiges of EU nationalism was farm policy.[14] We were provided with a way forward and wished great success. The critical piece here was that we now understood the regulatory and permitting process for our project in Belgium.

So, we thought.

The next day, Juan and I were scheduled to board a hovercraft to Dover (UK).[15] It was a Saturday. We had a coffee, a pastry, and then drove to Calais (France) and scheduled a short ride over English Channel. Calais is a magnet for weekenders looking for cheap goods in a duty-free zone. So, there were tourists shopping and businesspeople heading home, and the craft looked like it would be packed. That was not the concern. My concern was that a hovercraft is like a floating tube that slides a few feet off the water.

I just knew it was dangerous.

Decades earlier, the hovercraft top sized and 57 people were killed. Plus, I get very motion sick. On planes, in cars, and, especially on floating boats flying over the English Chanell. My recent surgery had made this condition worse. My guts were not symmetrical. The doctors had taken out almost 17 inches from my colon. I felt lopsided. Nausea required a comfortable, cool place to sit and think.

But that was not the real issue.

The real issue was that Juan had not gotten a VISA to enter the UK. As an Ecuadorian national, he needed permission from Her Majesty's Government to grace those isles. What could we do? D. Reggie may have been a hustler on the streets of Brussels, or the seller of fake phones, cigarettes, and whiskey in Lisbon, but I do not think he could spell Downing Street. Besides, we were on French soil. Since we had never met the UK team members, we did not feel comfortable reaching out to them. Besides, Juan should have taken care of this before we left.

"Let me call Ian," offered D. Reggie.

He meant Ian Selby. That was one of his partners, and he was a British National. Selby had made a fortune in the 1980s and was connected to people. Ian Selby had been roped into this project by a man named George Miller. Miller had worked for Prime Minister Margaret Thatcher as an agricultural policy expert. He had also conducted international business in Russia, South America, and throughout the Middle East. He could fix this. He had a friend (Danny Streetman) that had been in a Tory in Parliament representing a southern England region. Between them, they could manage this.[16]

Within an hour, the Port Authority at Calais received a fax from the British government insisting that Juan Josse was working closely with the Department for Environment, Food & Rural Affairs (DEFRA), like the USDA with the Food and Drug Administration (FDA) together.[17] And, as such, he was needed in the UK immediately. Please extend all courtesies, etc.

It worked.

It was on official (looking) letterhead. It might have been legitimate.

We were impressed.

A bit more relaxed now, Juan and I bought some tax-free wine and cheese, boarded the flying tube death-trap, and headed to Dover. It was largely uneventful. That did not prevent me from being nervous the entire trip. D. Reggie offered me a cigarette. I probably would have vomited if I had smoked one.

Our destination was not London. We first landed in Dover, cleared customs and immigration (Juan had his "letter" in hand), and found a rental car waiting for us. We travelled about 4 hours to the west to Dorchester County. Our end point was Piddlehinton, a village of about 400 people a few miles north.[18] This was a picturesque area of cathedrals, bubbling brooks, and thatch-roofed inns and houses. Thomas Hardy lived in this area and wrote about this place. It was what I though English country life was like. Maybe it was and still is.

We settled in and met more of "the boys" for an early dinner. They were an assortment of characters unlike anything I had ever seen. Juan and I were the youngest. Mostly, they were early 40s to early 80s. The common thread seemed to be that none of these people had "day jobs." Mostly, the boys did deals and moved on.

English humor is not for everyone. I grew up watching Monte Python, the Two Ronnies, and loved the Benny Hill Show. Later, I would discover Bernhard Manning. So, when I learned the origins of the name of the village, I fell out laughing. "To piddle" is a way of saying urinate. We were in a village named after having a pee while talking about hog manure. Irony? Not everyone in the restaurant that evening found this amusing. It would not be the last time we were chastised for talking crap at the meal table. Some people are so sensitive.

Ian Selby was a Cockney-talking businessman with deep pockets and well-heeled friends. He resembled a British TV star and was Jewish. Selby was in his early 50s. He had a look of perpetually smelling rotten eggs and it made him ideal for our project. He drove a powder blue Bentley with creamy white interior. He vacationed in Miami and loved women in bikinis and fast money. Ian was a likeable rascal.

Danny Streetman was a conservative politician, farmer, husband, and impossible to like. He was nasally, often congested, and always seemed to be blowing his nose. This made him speak in a muffled, yet whiny tone. He really disliked Americans and took every shot he could at me. He had worked and lived in Argentina and believed that he was a Gaucho. Yet, he was honest. I always respected him but could never like someone so insecure. He treated everyone like a servant that was stealing the silverware. Juan really liked Danny. He trusted him and believed him. I should have known that this would be the beginning of the end for Juan and me.

The enigma of the group was an American lawyer named Jim Tryand. Jim was short, round, missing several teeth, almost 70 years old when we met. He seemed much older. Jim had left his family one Christmas Eve decades earlier. He flew to Geneva and started a new life. It would be years before his family knew what happened to him. Jim was a graduate of Michigan Law School and had meandered the international legal waters for companies and interests all over the world. His occupation was listed as economist and lawyer. Jim looked like an old boxer with a flat nose, and even flatter ears. He was not punchy. Jim was incredibly sharp witted. Yet, he was one of the more dangerous people I have ever met. He seemed disarmingly helpless and vulnerable. Neither of these were true. He was a survivor with stunning attributes for somehow landing on his feet. Jim was like an old cat. The one that outlives all the tougher, meaner dogs. The fact that he could walk away from his family, especially his son, told me that he was to be feared. A person that could do that was capable of anything.

George Miller was the leader. He was always the leader. He had a great heart and a huge smile. Typically, he wore a neatly trimmed beard and mustache. His education was superior. George's reputation was spotless. He was courageous and brave to the point of appearing to be reckless. He did not believe in fear. He was Russian Orthodox, married to a German, and had two teenage kids. He was a vehement anti-Communist but loved the British Monarchy. George was an Adam Smith Tory. He spoke several languages, loved good food and wine, and was clever and refined. He was the opposite of Ian Selby in almost every way imaginable.

I admired George Miller and aspired to be him one day. A cultured, cosmopolitan, dealmaker, with a passion to make a difference. George was never predominantly motivated by money. He wanted to change lives. Maybe it was my religious background, but I felt we had an obligation to use our talents to make the world better. Over the next several years, George would give me books and articles that pressed me to justify my beliefs. He challenged me to think in ways that I had never done.

The evening was filled with stories about deals that had been successful and those that went horribly wrong. Jim Tryand, for example, had brokered a gold deal with a Nigerian bank back in the late 1970s. His client had paid for the gold, but the gold never showed up. When he went to Lagos, it was revealed that the "vice-president of the bank" was a janitor that held meetings in the VP's office during extended lunch breaks. The janitor turned VP had changed clothes and even had a phony name plate on the door and fake business cards. His "secretary" was a hotel maid from across the street. He had scammed dozens of foreigners out of their money before fleeing Nigeria. Tryand found the guy a few years later in Vienna. That is where the story ended.

Silence.

Then, George Miller talked about the time he negotiated the sale of farming equipment on the outskirts of Moscow during a harsh Russian winter. His "partner" and he had their left hands tied together and were given pistols and forced to "resolve their conflicts" or someone else would. Miller just grabbed the gun and fired point-blank range in the man's face. Click. Click. Click. The gun was not loaded. It had been a test. You had better be certain that you are right before you pull the trigger. George completed the deal, sold a ton of tractors, and made a lot of money. Everyone laughed and toasted to success, "Cheers!"

I sat amazed.

Years later, in Abu Dhabi, I encountered a similar situation. No guns, just knives.

Back when I enjoyed a beverage or three, I loved dark English beers. Room temperature. Maybe it was the Samuel Smith Oatmeal Stout effecting my judgement, but I was impressed. A little scared, too. Still, these people could take us to the next level. Sure, we would have to watch D. Reggie and, for certain, Tryand. But the deal could work. Miller was the real deal.

We retired for the evening and agreed to meet for breakfast the following morning. Not too early. Juan and I spoke after dinner in my room.

"What do you think?" I inquired.

"I really think we can work with Danny."

"Really?" I was astonished.

"He knows the industry and works. The rest of these people don't seem to have any job."

He was right about that. They lived off deals. Maybe that was just the type we needed to make this project succeed.

The next question made me assess our relationship for the first time.

"What role to do you see for yourself?" Juan posed.

"What do you mean?"

I was confused and a bit worried. What was going on here?

"I think they want me to move here in the south of England or to Belgium. What role do you think you will play from now on?" He was serious.

I had assumed that Juan and I were tied together. Brothers in research. By this point we were busily working on the first patent to protect our intellectual property (IP) rights. I did not think he would ever consider abandoning me and all our work together. I had worked in the heat, covered in sweat, and waste to make this occur. I had almost died in North Carolina. Was it not me that brought these people to the table?

What did this mean?

Juan was going to cut me out of my own deal.

So, he thought.

I had no intention of allowing that. From that day forward, Juan and I were never the same. We worked together and spent recreational time together, but I could never trust him 100%. I was very hurt. Deeply. Juan was like my brother. But Able had a brother, too.

Before the other arose the next morning, I took a long walk along the cool stream and flowered roadside. I thought. I thought a lot. I had to prove my value, my worth. I began to throw myself into all EU[19] and UK[20] policies and regulations, along with new research. There was some interesting work from Scotland that I began to review.[21]

Once again, Asian states were doing some foreword thinking projects to reduce the ecological scourge of pig waste on water and air.[22] I needed to become more of an expert than my expert colleague. I could never match him for water engineering. However, I had a good, practical mind and I understood cost-benefit analysis. Also, I could wade through laws and comprehend their intent. By combining my limited engineering knowledge and experience, I could prove my worth. I could also sort through the agriculture literature to understand what farmers needed to stay in business.[23]

I needed to meet with George Miller and have a candid conversation.

We would meet in London later the following month.

Just George Miller and me.

None of the boys.

And no Juan Josse.

It would be the first time I met his family, stayed in his home, and enjoyed a long discussion about economics, politics, and philosophy. He beat me in chess several times. He introduced me to several books that offered an insight into another world.

It would not be the last time we spent time together at his house.

I became a permanent guest and had my own room.

His family extended every possible courtesy.

Once he opened his home to my wife and I to stay there when we had tickets to see Kiss Me Kate at the Victoria Palace Theatre in London.

George Miller was a remarkable person. A true friend.

NOTES

1. Tive, C. (2006). *419 scam: Exploits of the Nigerian con man.* iUniverse.

2. Smith, A. (2009). Nigerian scam e-mails and the charms of capital. *Cultural Studies, 23*(1), 27–47.

3. You can't cheat an honest man. (1939). Located at https://www.imdb.com/title/tt0032152/. Accessed 11 June 2022.

4. Bhattacharya, C. (2001). Intellectual property rights: Violations in China. Retrieved February, 14, 2006.

5. Teodoro, J. A. R., Pereira, H. V., Sena, M. M., Piccin, E., Zacca, J. J., & Augusti, R. (2017). Paper spray mass spectrometry and chemometric tools for a fast and reliable identification of counterfeit blended Scottish whiskies. *Food Chemistry, 237*, 1058–1064.

6. Ostende. Located at https://www.visitoostende.be/en. Accessed 11 June 2022.

7. Focus on Belgium. (2019, October 21). Did you know that Marvin Gaye loved in Ostende? Located at https://focusonbelgium.be/en/facts/did-you-know-marvin-gaye-lived-ostend. 2 July 2022.

8. *South China Morning Post.* (2020, September 5). Why Ostend, the Belgian town where Marvin Gaye cleaned up, is the creative birthplace of Sexual Healing. Located at https://www.scmp.com/lifestyle/travel-leisure/article/3099870/why-ostend-belgian-town-where-marvin-gaye-cleaned-creative. Accessed 2 July 2022.

9. Pratt, S. J., Frarey, L., & Carr, A. (1997). A comparison of US and UK law regarding pollution from agricultural runoff. *Drake L. Rev., 45*, 159.

10. Munroe, J. A. (1995). Sitting livestock and poultry operations for the 21st Century-Symposium proceedings, Ottawa, Ontario, July 13–14, 1995.

11. Seidl, A., & Grannis, J. (1998). Swine policy decision points. Agricultural and resource policy report (Colorado State University, Dept. of Agricultural and Resource Economics); ARPR 98–01.

12. European Commission. Located at https://ec.europa.eu/info/food-farming-fisheries_en. Accessed 11 July 2022. Today, much of their efforts focus on organic foods, sustainable agriculture, and climate change mitigation and adaptation policies. Managing CAFOs was still a hot topic when we arrived. Experts from EU universities and other research facilities were doing extraordinary work.

13. Baylis, K., Peplow, S., Rausser, G., & Simon, L. (2008). Agri-environmental policies in the EU and United States: A comparison. *Ecological Economics, 65*(4), 753–764. Another excellent work includes the following: Stolze, M., Piorr, A., Häring, A. M., & Dabbert, S. (2000). *Environmental impacts of organic farming in Europe.* Universität Hohenheim, Stuttgart-Hohenheim. Their commitment to

preserving small, family farms began more than a decade before such efforts in the US. Potter, C., & Lobley, M. (1993). Helping small farms and keeping Europe beautiful: A critical review of the environmental case for supporting the small family farm. *Land Use Policy, 10*(4), 267–279.

14. Happe, K. (2004). Agricultural policies and farm structures-Agent-based modelling and application to EU-policy reform (No. 920-2016-72838). There is a recent work on this subject for the UK. Carter, A. (2020). Flags don't feed people: Nationalism and agriculture in the UK: A critical framing analysis of nationalism as it appears in British agricultural discourse in the context of Brexit. [MSc thesis].

15. BBC News. (2022, March 4). Hovercraft capsize disaster off Hampshire coast recalled 50 years on. Located at https://www.bbc.com/news/uk-england-hampshire -60236843. Accessed 5 June 2022.

16. His name has been changed to protect his family's privacy.

17. Department for Environment Food & Rural Affairs. Located at https://www .gov.uk/government/organisations/department-for-environment-food-rural-affairs. Accessed 13 May 2022.

18. Visit Dorset. Located at https://www.visit-dorset.com/listing/piddletrenthide -and-piddlehinton/112430301/. Accessed 7 June 2022.

19. Windhorst, H. W. (1998). Sectoral and regional patterns of pig production in the EU. *Pig News and Information, 19*(1), 232–249.

20. House, M., & Pool, K. (1990). Food and Environment Protection Act, 1985, Part III.

21. Hillman, J. R. (1999). Report of the director. Scottish Crop Research Institute, 11.

22. Taiganides, E. P. (1992). Design and construction of the Ponggol pig waste plant. Pig waste management and recycling; the Singapore experience, 263–358.

23. Hobbs, P. J., Misselbrook, T. H., & Cumby, T. R. (1999). Production and emission of odours and gases from ageing pig waste. *Journal of Agricultural Engineering Research, 72*(3), 291–298.

Chapter 7

A Piggy Patent, Part I

The trip to Belgium and the UK was behind us now. We had travelled safely back home. There would be some family time and then back to work. Back to the Swine Unit. But we had other demands. Now we had to complete the patent to secure our seat at the table with the boys. Neither of us had ever written a patent. Patent law is notoriously difficult and the methods of describing technical claims are essential to establishing a commercial value for the intellectual property (IP).[1] I must have read a book or an article about start-up companies needing IP to create value for their business.[2] Everyone tends to believe that. To pursue the patent unicorn in hope that the myth becomes a reality. It is a myth probably started by patent attorneys.

Within patent law, there are many nuances and specializations. All the engineering subfields have applicable patent processes. Chemical engineering presented itself as the most relevant and the most difficult.[3] We needed to start with the basics.[4] Then, we needed to grasp the subtilties of European Union (EU), the UK, and American intellectual property (IP) systems.[5] And if we were ever to enter the Asian market, knowing something about Japanese law would be useful.[6]

The IP here, though, covered more than chemical engineering. It included agriculture considerations. IP for farming added another layer of complexity.[7] At an international meeting of the American Society of Agricultural Engineers in Florida (1998), there was considerable discussions regarding CAFOs and pig waste.[8] All seemed to agree: anerobic treatment was the only way forward. Trying to feed oxygen into the slurry after it drained into the lagoon was capita intensive and would not work. We knew we were on the right track, but it is always good to have confirmation from real experts. In fact, for almost 20 years (at that time), scientists had agreed on this point.[9] And not just for pigs. Waste from cows and chickens must be managed anaerobically.[10] The issue was not whether it was anaerobic or not. The question always seemed to be one of scale. There were tons of research reports, published articles, and demonstration projects that established anaerobic treatment options as best.[11] But adoption of a new technology at a CAFO remained allusive.

Why?

Even in a demonstration scale model, like the one we built on the back of a trailer, everything had to be oversized. All the reactors, pumps, and piping system had to be scaled for a CAFO even though we were treating waste from only 300 hogs. Why? Because the urine and feces are still the same size. The volume is different. But the physical size of the waste slurry is the same. We were not treating waste from Guinea pigs. These were large hogs. To be fair, problems with scale for any ecological research is always a worry.[12] So everything was oversized for research. That was issue number one: finding the balance between treatment and cost for a CAFO.

Next, we knew that accounting for pig pee and poop was essential. Yet we had no idea that pig sweat would become so serious. Scientists had long established that pigs do not sweat in the same manner that humans or other mammals do.[13] They sweat largely from their noses. As humans sweat, we lose fluids and salt (sodium chloride; NaCL). To maintain an internal temperature, pH, and salt balance, pigs had to 'sweat' in another way. That NaCL loss came through their waste. When pigs urinate, their pee contains phosphate (PO4), a type of salt compound, and ammonium (NH4). The chemistry is simple (now that I know it). When magnesium (Mg) is added to urine, phosphate, ammonium, and magnesium bind and form struvite ($MgNH4PO4 \cdot 6H2O$).[14] There has been considerable effort in trying to extract that struvite from pig slurry or lagoons.[15]

Down at the Swine Unit, we discovered struvite in an unorthodox manner. Several of the pumps had burned out and most the piping was so clogged that there was no flow. As the flow backed up into the reactors, the system would overflow, and waste covered the trailer. Washing down the trailer became a morning ritual. Then, we decided to pressure wash the insides of the piping to clear any debris or large solids. Pigs lose a hair all the time and especially in the summer.[16] As expected, there was pig hair throughout the piping and even in the pumping system. The pump membranes and seals between pipe joints were cracked. Pig hair is abrasive.

"Don," I said to one of our worker, "Could you climb up to the pipes leading from the holding tanks to the reactors? Then, test if there is any pressure?"

I wanted to see where any blockage had begun. Was it close to the reactors? Closer to the holding tanks?

"You got it, boss man," Don said. He knew that I hated his sarcasm. He thought I was some young, smart aleck born into this world to make his life miserable.

"Be careful."

"Boss man, I have been climbing ladders, testing pipes, and cutting pipes, before your momma even knew your daddy," he reminded he, "I think I can manage."

"Wayne," I asked, "Could you help Don so he does not kill himself?" Wayne was younger and a little crazy. He once asked permission to cut the Swine Unit's manager for insulting me. He said that he was wanted in six states and could disappear before it got dark. I told him thanks and that I would keep that in mind for later. In the meanwhile, I needed him to assist Don.

Don began smacking the pipe with his 20-ounce Estwing hammer.

"It   sounds   solid."

"Drill a small hole in the side and relieve some of the pressure."

"I know how to test the pressure on a pipe, son!"

Don proceeded to cut through the pipe using a fine-toothed saw. He cut through about an inch when the pressure burst through. His face was directly in front of the release point. He was covered in pig hair, urine, feces, and struvite. He fell backwards off the ladder and onto his back. He was not hurt too bad. He was startled and embarrassed.

I stepped over him and said, "Well, break-time is over, Don. Get up."

He mumbled some words about the questionable nature of my mother, my father, and told me to be fruitful and multiply myself. Not exactly like that. But you get the point.

This was a 4-inch pipe completely packed with struvite for about 6 feet. There was little we could do with this other than dispose of it. Maybe it had some nutrient value or could be part of a total waste elimination system.[17] For now, it was a nuisance. This meant we needed to filter the water better and prepare for additional solids treatment.

By this point, the notion of including any solids treatment in the IP was eliminated. Each facility would have its own needs for solids. Maybe the solids would be recycled for food or for their nutrients.[18] Even using the waste for nutrient uptake with plants in the lagoon was a thought. Duckweed research then and now is still a good idea.[19] I have always preferred eucalyptus because it sucks up so much water and requires a lot of nutrients and it has a pleasant aroma.[20] Reuse and recycling of pig waste for biomass to biogas has been a CAFO dream forever.[21] In high-energy country of the EU, like the UK, this would be supported by the farming industry and the government.[22] Further, this had the advantage of using much of the water fraction of the waste stream as well.[23]

If we were going to build a facility in northern Belgium, we needed to focus on creating a fuel from the solid fraction. The idea of dewatering the sludge, adding polymer, and extruding energy pellets for industrial boilers was the step forward.[24] We could access other waste streams and comingle the sludge to produce a fuel-pellet. The obvious place to start was with sewage sludge.[25] While sewage sludge must be dewatered and it has a low- or medium-BTU value, the mixing of waste streams could create a homogenized pellet that could consistently fire up an industrial boiler.[26]

Typically, submitting a patent requires a lawyer, lots of money, and time. We could manage one of these, but not all three. Juan and I had brains. We did not have thousands of dollars and months to spend. Maybe we could do it? The process for writing a patent requires much pre-production work. An outline of where you want to go. You must know what makes your invention unique. What is the prior art?[27] Who filed previous patents like this? Did that patent become commercially available?[28] The measure of success for patents is not the novelty but the adoption and its commercial success.[29] Just because no one has ever invented something does not necessarily mean it has value. Often, the cost of patenting an invention is higher than the economic value of it.[30]

We believed that we had something unique, and it had value.

Patent law has its origins in the adoption of the US Constitution. There are many pros and cons with this process. In short, the founders stated, "Congress shall have power . . . To promote the progress of science and useful arts, by securing for limited times to authors and inventors the exclusive right to their respective writings and discoveries."[31] Decades of court battles attempting to offer clarification of intent and authority of the US Patent and Trademark Office (USPTO), Congress passed the US *Patent Act* (found in Title 35 of the United States Code). It would be amended by the *Leahy-Smith America Invents Act* (2011). But in the early 2000s, we were still under the old system: first to file gets the patent.[32]

According to the law, a person shall be entitled to a patent unless, "[T]he invention was known or used by others in this country, or patented or described in a printed publication in this or a foreign country, before the invention thereof by the applicant for patent . . . "[33] Thus, we were in a race with anyone else. Maybe even our old friend Jordan Hill?

The claims made by the patent application are the true value.[34] According to the Legal Information Institute, "A patent claim defines the boundaries of an invention, and therefore lays down what the patent does and does not cover. A patent claim is the most important thing in a patent application, for it defines the subject matter that is sought to be protected. And the rest of patent specifications are explaining the invention in detail. A patent claim is usually expressed as a statement of technical facts in legal terms."[35] The USPTO provides guidance for patent submissions.[36]

Next, we had to figure out what type of patent to file? Would it be a device or apparatus? Or would it be a process? Again, we consulted with the

USPTO.[37] We opted for both. Two patents were filed within a couple of years of one another. The first was the apparatus patent.

This patent was granted in 2002 and was named the 'Fluidized Bed Reactor Apparatus.' The design was really very simple . . . eventually. Yet, this is the most critical breakthrough of all the research. A single cone-shaped bottom to facilitate the even distribution of the bacteria to expedite the breakdown of nitrogen and phosphorus. In words, this made it possible to eliminate the ammonia from the water.

In more technical terms, according to the patent, it was "A biological fluidized bed apparatus for the treatment of wastewater comprises: a reactor tank having a conical bottom with angles ranging from about 30 to about 80 degrees, and most preferably about 60 degrees from a horizontal plane; an inert media bed within said tank; and a recirculation pipe within said tank having an upper inlet, and a lower conformed outlet fitting within said media. The apparatus is self-contained and offers easy and uniform expansion for various types of commercially available media. Besides the biological treatment accomplished within the fluidized bed, this reactor provides mechanical solids/gas/liquid separation and up flow sludge blanket clarification to improve effluent quality. Preferred additional features of the apparatus include a rotating surface skimmer and an excess sludge collection and thickening device powered by an externally mounted drive. With this reactor configuration, anaerobic and anoxic biological treatment and subsequent clarification are achieved in a single enclosed tank."[38]

Again, this was the first of the patents filed.

The next patent would take another 18 months to complete. It is hard to believe what happened over the next year and a half. Lots of trips to the UK and northern Europe.

Countless meetings. Way too many presentations.

Astonishing and outrageous experiences.

And some interesting false starts that took up time and resources. One of the more insane flirtations was the quest to make the device energy efficient. Here, we attempted to capture the biogas produced from the decomposition of the solids, scrub the gas of moisture, and produce methane. The idea was that the methane would power at least part of the system and off-set any electrical power demands.[39]

It worked.

We captured the biogas.

We scrubbed the moisture.

We had methane.

The capital costs for doing this exceeded the energy savings by a large margin.

This was another example of how cool things can happen in research that have no practical value on a full-scale system.

Our device was never going to be the perpetual motion machine we envisioned.

But it was going to make a lot of money.

NOTES

1. Burk, D. L., & Lemley, M. A. (2002). Is patent law technology-specific? *Berkeley Technical Law Journal, 17*, 1155.

2. Feldman, R. (2013). Patent demands & startup companies: The view from the venture capital community. *Yale JL & Tech., 16*, 236.

3. Breyer, W. S. (1993). Know the basics to protect your inventions. *Chemical Engineering, 100*(9), 120.

4. Rosenberg, P. D. (1992). *Patent law basics*. New York: Clark Boardman Callaghan.

5. Van Overwalle, G. (1997). *The legal protection of biotechnological inventions in Europe and in the United States: Current framework and future developments*. Another excellent source. DA, C. P. (1998). Legal protection of biotechnological inventions. The European Parliament's perspective. *Revista de Derecho y Genoma Humano= Law and the Human Genome Review, 8*, 71–103.

6. Schütt, C. (2004). Patents for biotechnological inventions: Current legal situation and case law in Europe, the US and Japan (Master's thesis, ETH, Eidgenössische Technische Hochschule Zürich, Professur für Rechtswissenschaft). There is not real protection under Chinese IP law.

7. Foltz, J., Barham, B., & Kim, K. (2000). Universities and agricultural biotechnology patent production. *Agribusiness: An International Journal, 16*(1), 82–95.

8. Chynoweth, D. P., Wilkie, A. C., & Owens, J. M. (1998). Anaerobic processing of piggery wastes: A review. In ASAE Annual International Meeting, Orlando, Florida, USA, 12–16 July, 1998. American Society of Agricultural Engineers (ASAE).

9. Velsen, A. F. M. V. (1981). Anaerobic digestion of piggery waste (Doctoral dissertation, Landbouwhogeschool te Wageningen).

10. Pain, B. F., & Hepherd, R. Q. (1985). Anaerobic digestion of livestock wastes. In Anaerobic digestion of farm waste. Proceedings of meeting, NIRD, Reading, UK, July 1983 (pp. 9–14).

11. Vetter, R. L., Friederick, D. J., & Huntington, P. (1990). Full scale anaerobic digester and waste management system for a 300-cow dairy. In Agricultural and food processing waste: proceedings of the 6th International Symposium on agricultural and food processing wastes, December 17–18, 1990, Chicago, USA (pp. 236–249). American Society of Agricultural Engineers.

12. Levin, S. A. (1992). The problem of pattern and scale in ecology: The Robert H. MacArthur award lecture. *Ecology, 73*(6), 1943–1967.

13. Ingram, D. L. (1965). Evaporative cooling in the pig. *Nature, 207*(4995), 415–416.

14. Wrigley, T. J., Webb, K. M., & Venkitachalm, H. (1992). A laboratory study of struvite precipitation after anaerobic digestion of piggery wastes. Bioresource technology, 41(2), 117–121.

15. Beal, L. J., Burns, R. T., & Stalder, K. J. (1999, July). Effect of anaerobic digestion on struvite production for nutrient removal from swine waste prior to land application. In ASAE Annual International Meeting. Paper (No. 994042).

16. Bustamante, M. A., Moral, R., Bonmatí, A., Palatsí, J., Solé-Mauri, F., & Bernal, M. P. (2014). Integrated waste management combining anaerobic and aerobic treatment: A case study. *Waste and Biomass Valorization, 5*(3), 481–490.

17. Kalyuzhnyi, S., Sklyar, V., Rodriguez-Martinez, J., Archipchenko, I., Barboulina, I., Orlova, O., . . . & Klapwijk, A. (2000). Integrated mechanical, biological and physico-chemical treatment of liquid manure streams. *Water Science and Technology, 41*(12), 175–182.

18. Aarnink, A. J. A., & Verstegen, M. W. A. (2007). Nutrition, key factor to reduce environmental load from pig production. *Livestock Science, 109*(1–3), 194–203.

19. Gwaze, F. R., & Mwale, M. (2015). The prospect of duckweed in pig nutrition: A review. *Journal of Agricultural Science, 7*(11), 189.

20. Higashikawa, F. S., Silva, C. A., Nunes, C. A., Bettiol, W., & Guerreiro, M. C. (2016). Physico-chemical evaluation of organic wastes compost-based substrates for Eucalyptus seedlings growth. Communications in Soil Science and Plant Analysis, 47(5), 581–592.

21. Arkhipchenko, I. A. (1998). Microbial fertilizers from pig farm wastes. *Ramiran, 98*, 439.

22. Griffiths, A. J., & Hicks, W. (1997). Agricultural waste to energy—a UK perspective. *Energy & Environment, 8*(2), 151–167.

23. Marr, J. B., & Facey, R. M. (1995). Agricultural waste. *Water Environment Research, 67*(4), 503–507.

24. Griffiths and Hicks.

25. Have, H., & Henriksen, K. S. (1998). An Energy-Efficient Combustion System for High-Moisture Organic Wastes and Biomasses. *Water and Environment Journal, 12*(3), 224–232.

26. Klass, D. L. (1985). Energy from biomass and wastes: A review and 1983 update. *Resources and conservation, 11*(3–4), 157–239. A more recent article. Pytlar Jr, T. S. (2010, January). Status of existing biomass gasification and pyrolysis facilities in North America. In North American Waste-to-Energy Conference (Vol. 43932, pp. 141–154).

27. Callaert, J., Van Looy, B., Verbeek, A., Debackere, K., & Thijs, B. (2006). Traces of prior art: An analysis of non-patent references found in patent documents. *Scientometrics, 69*(1), 3–20.

28. Merges, R. P. (1988). Commercial success and patent standards: Economic perspectives on innovation. *Calif. L. Rev., 76*, 803.

29. McAleer, M., & Slottje, D. (2005). A new measure of innovation: The patent success ratio. *Scientometrics, 63*(3), 421–429.

30. Mazzoleni, R., & Nelson, R. R. (1998). The benefits and costs of strong patent protection: a contribution to the current debate. *Research Policy, 27*(3), 273–284.

31. US Constitution. Article I, Section 8.

32. 35 U.S.C. 100. This would be modified later.

33. 35 U.S.C. 102. Additional provisions include: "(b) the invention was patented or described in a printed publication in this or a foreign country or in public use or on sale in this country, more than one year prior to the date of the application for patent in the United States, or (c) he has abandoned the invention, or (d) the invention was first patented or caused to be patented, or was the subject of an inventor's certificate, by the applicant or his legal representatives or assigns in a foreign country prior to the date of the application for patent in this country on an application for patent or inventor's certificate filed more than twelve months before the filing of the application in the United States . . ."

34. Marco, A. C., Sarnoff, J. D., & Charles, A. W. (2019). Patent claims and patent scope. *Research Policy, 48*(9), 103790.

35. Legal Information Institute. Located at https://www.law.cornell.edu/wex/patent _claim. Accessed 23 April 2022.

36. US Patent and Trademark Office (USPTO). Located at https://www.uspto.gov/ web/offices/pac/mpep/s1824.html. Accessed 23 April 2022.

37. USPTO. Located at https://www.uspto.gov/web/offices/pac/mpep/s2114.html. Accessed 25 April 2022.

38. Fluidized Bed Reactor Apparatus. Located at https://patents.google.com/patent /US20030209476A1/en. Accessed 4 January 2022. The official patent number is US20030209476A1.

39. Price, E. C., & Cheremisinoff, P. N. (1981). *Biogas: Production and utilization.* Ann Arbor. Another great source that is more current. Weiland, P. (2010). Biogas production: Current state and perspectives. *Applied Microbiology and Biotechnology, 85*(4), 849–860.

Chapter 8

The Morgenthau Notes

People love treasure hunts. We got suckered into a treasure hunt by a modern-day pirate. He did not have a Jolly Roger waving above his ragged ship. But a pirate he was. First some context would be worthwhile. Be patient. This gets insane.

On March 9, 1933, President Franklin D. Roosevelt through Executive Order 6102 made it illegal for citizens to own gold.[1] Speculation over how the American government spent that gold has been endless. At the center of all controversy was one of Roosevelt's closet ally Henry Morgenthau.[2] These two had been neighbors, friends, and public servants for many years. According to a PBS documentary, "When Roosevelt became governor of New York in 1928, he appointed Morgenthau the chairman of his agricultural advisory commission. When Roosevelt was elected President in 1932, Morgenthau became his Treasury Secretary."[3]

Morgenthau's road to Washington, DC was not foreshadowed by a political career. According to Federal Reserve documents, "He studied architecture and agriculture for two years at Cornell University before dropping out in 1913 to become a farmer. He bought 1,000 acres of land in Duchess County, New York, and it was at this time that he first met Franklin and Eleanor Roosevelt. In 1922, Morgenthau bought the American Agriculturist magazine."[4] His succession to Treasury was preceded by Roosevelt's original choice for that position becoming ill.

But his influence extended past New York or Washington, DC, politics. "Morgenthau's most significant and lasting impact on the world economy came at the Bretton Woods Conference in New Hampshire in 1944, where he served as chairman. The conference established the International Monetary Fund and the International Bank for Reconstruction and Development and pegged all international currencies to the dollar."[5] Morgenthau published his justification for American leadership through the auspices of the United Nations system in the prestigious Foreign Affairs journal firmly establishing the dominance of the dollar for decades.[6] Upon leaving government service,

Henry Morgenthau dedicated his life to philanthropic causes, especially those benefitting Jewish charities.[7]

His was an extraordinary life.

So how does this connect with the quest to cleanup pig poop?

Great question!

Here comes a dismal and astonishing answer.

And it is true.

By the third or fourth trip to the UK and northern Belgium, it had become clear that Juan and I would be spending more time overseas than on the Swine Unit in Baton Rouge. For me, that meant my family, which would soon include another baby, could move back to New Orleans. After several weeks of house-hunting, we found a wonderful shotgun house on historic Bayou St. John in the heart of New Orleans.[8] It was central to both of my offices at the University of New Orleans Urban Waste Management & Research Center (UWMRC) and Tulane University. Also, and this was very important, it was walking distance from the Fair Grounds, which doubled as a world renown spring music festival and a thoroughbred racing track.[9] I love music. I love the ponies. The house was also walking distance from our church, coffee shop, and grocery store. The Edgar Degas home was within two blocks.[10] The Old Spanish Custom House was across the street.[11]

It remains my favorite place I ever lived in New Orleans.

After settling into my new home, adjusting to my new teaching schedule, and working on another patent, I was headed back "across the pond." George Miller had important news and could not discuss it on an international telephone line. This became the *modus operandi* for Miller, especially after 9/11. That was still yet to happen.

"Johnny baby," George loved to call me this, and I came to like it as well, "You yanks just went through your most turbulent election ever. There will be uncertainty as you have never seen. America has shown that it has cracks in her Statue of Liberty. And you have built an industry of enemies looking to kick you for any reason."

This was not the last time George's words would prove preternatural.

If that has come from Danny Streetman, a committed Anti-American, I would have probably given him a smack in the mouth. Coming from George Miller, I took it seriously. The turbulence he referenced was the 2000 Election and uncertainty was to follow closely.[12]

"But we have business, and we won't allow American domestic unrest to delay our efforts, right?" he added.

I agreed.

We needed to press forward.

By this time, the ravages of Hurricane Floyd were being fully understood. Belgian, Dutch, and German pig farmers were looking for alternatives. American CAFOs were feeling pressure. The time was now.

George had used his influence to gain access to the UK Department for Environment, Food and Rural Affairs (DEFRA).[13] He had coordinated an economic analysis with Streetman and Jim Tryand with input from Juan and me. We met at the National Liberal Club in London with various agricultural experts and government agents. The National Liberal Club stands on Whitehall as a bastion of progressive politics and has one of the most beautiful spiral staircases in the world.[14] Tradition coupled with refine elegance greets you at the door. A delicious lunch followed by an afternoon tea with a quiet cigar and a brandy before going home.

This was how a business meeting is supposed to be held.

An interesting note about the meeting. George led the meeting. Not as a dictator, but it was clear who was in-charge. Jim Tryand, the economist-lawyer, sat in the corner and took notes. I learned that day how different British and American lawyers were expected to engage in meetings.[15] He never said a word unless he was asked a question by George. Danny added some context every so often. He wanted everyone to know that he was a real farmer. He had ridden horses and rope cows in Argentina after all. Juan and I participated on technology discussions. Mainly we demonstrated our competence in water technology. Ian Selby provided little as his expertise would be tested after we were funded. He seemed bored. D. Reggie and Shaky Eddie were not invited. They would be briefed later if at all.

The primary inferences evolved slowly. First, the solids research would need to be more flexible and would have to adapt to local circumstances, such as access to other waste streams. Second, there was no need to build another demonstration sized system. The concept had been proven. We needed to move ahead with a full-scale system handling the waste of a large CAFO or several smaller CAFOs. Third, we had a meeting later that week with an investor prepared to support the project. This was what sparked the need to meet in person.

George Miller had important friends. Wealthy, well-connected political friends. One of those was a descendant of the McWirter Family. In the 1950s, following a dispute regarding bird trivia, twin brothers decided they would research and write down all the answers to those crazy questions that people ask when sharing a pint. Who is the tallest man? Who ate the most hotdogs? Where is the lowest point on Earth? Today, the Guinness Book of World Records is universally recognized as the source for settling (these critical) disputes.[16] Strangely enough, the original question involving the fastest gamebird may have never been resolved.[17]

Through an introduction from McWirter, George had met an international investor. His name was Nicholas Bristow. Bristow was slightly taller than six feet, had a giant head, an enormous neck, and an immense body. Imagine Ignatius Riley clean-shaven.[18] Gravity seemed to favor his shape; otherwise, he would have tilted over to one side. He was Canadian, married to a Philippine woman (all "the boys" called her "Ting Tong" after some British TV show[19]), and weighed more than 500 pounds. Everyone called him "Blobby." I never could. Of course, he knew what he looked like. I was not going to make an issue out of his weight and health problems. He mainly wore shorts, and his long shirts were always untucked. He claimed to be a Federal Reserve Agent.[20] This may or may not be a legitimate position within the Federal Reserve.

"Most people have no idea about their history," Bristow declared, "If they understood what World War Two was about, then they would ask more questions. They would try to understand the redistribution of wealth that took place world-wide. The Nazis were not just jackboot thugs, they were economic bandits."

He continued, "The world wars shifted power and sovereignty from empires and nations to international organizations. First was the Federal Reserve system in the US. Then, the United Nations, and the World Bank and the International Monetary Fund."[21]

There was more to his debatable lecture, "Bretton Woods was not just a vacation spot in New Hampshire. It realigned economic power around the dollar for decades."[22]

There was some truth to all these statements.

"And" he concluded, "Who was at the heart of all of this?"

We all sat silent.

"Morgenthau."

Henry Morgenthau's story lends itself to embellishment. He served in many capacities and was at the helm of the Treasury Department when American gold was made illegal.[23] There are so many allegations and myths and lies about Henry Morgenthau. And not enough actual scholarship to separate fantasy from reality.

"Ok," I inquired, "So what?"

"So what?" the question was spit out in disgust.

"The 'so what?' is that Morgenthau, a Jew, was trying to buy his brothers and sisters from Hitler using the gold he outlawed from Americans."

There are countless stories like this. There has been some research on this subject: Nazis selling Jews for money to wage war.[24] The Romanian Jewish story has some merit. There are also tales of Jews using property, art, gold, whatever it took to buy their way to friendlier countries.[25] There is no longer any disagreement over Nazis plundering Jews during World War Two.[26] I was familiar with all of this. I enjoy discussing history. Much of this was what we would call "counterfactuals" or alternatives to history. Interesting sources for debates.

Yet, what did any of this have to do with investing money in a pig poop invention in Belgium?

So, I asked, "What does any of this have to do with hog crap?"

George interjected, "Johnny baby, be patient." He nodded, to Bristow to continue.

I was not alone. Streetman and Selby were visibly aggravated. Tryand had, apparently, heard this story already. Juan and I were both dumbfounded and anxious.

Blobby rumbled on as we all sat breathless like kids around a campfire, "Morgenthau now had the means, the motivation. He just needed to act. Hitler also had gold. Gold that he had looted from Jews across Europe. A drop on gold prices would destroy his efforts. He meant to spread the risk around. Hitler required US currency. So, Hitler and Morgenthau exchanged Nazi gold for US Treasury notes, or bearer bonds. These notes were placed in boxes, nowadays called Morgenthau boxes or sometimes black boxes. But the plane from the Philippines to Germany never arrived. Something went wrong. Maybe it was the Japanese, maybe it was mechanical problems. The plane went down."[27]

A pause. He took a sip of tea.

Then, he continued, "There would have been maybe 100 of these black boxes. Each one with $100 million in bearer bonds. Even without any interest each box would be worth more than $2 billion today."

Dead silence.

The echo of silence gnawed at me.

What could all this mean? Was this real?

"About six months ago, the German government revealed that they had some of those boxes," Bristow avouched, "That their war-time ally Italy had shot down the plane. Confiscated the cargo and kept the loot for the past few decades. Others are threatening to come forward with these black boxes."

No one spoke.

Like his sagging flesh, Blobby let it hang until it drooped. His thick neck rolled as he swallowed and began the narrative again.

"As an agent with the Federal Reserve," he asseverated, "I have a duty to introduce, recirculate this money back into the global economy. There is a catch, of course. That money must support humanitarian causes. Environmental causes fit that requirement. Water is an area that I am very concerned. Dr. Sutherlin, he is recognized by the United Nations as an international water expert. Juan Josse is a water engineer. This project is a water project."

His incantations had worked.

We were spellbound.

Then the magic words.

The rotund sorcery finished, "Assuming that your technology is sound, and your cost estimates are correct, I stand ready to fully fund up to 10 of these projects."

There was no yelling or screaming or jumping up and down. This was the moment we had been working for and we now had a path forward. Blobby had come to George through venerable sources. The worthy McWirter name alone was like British royalty. Tryand claimed to have "checked Blobby out." Only Streetman and Selby had their doubts. They were businessmen. Like Thomas, though, we all wanted to see proof.

We would soon get that proof.

If Blobby had access to billions of dollars, then let's see some of it.

We did not have to wait much longer.

Within a few days, the boys, Juan and I and D. Reggie and Shaky Eddie were back in Ostende. This time, we were confirming the pig waste contracts with regional farmers along the Flemish coast and looking for potential sites to build the first facility. Blobby had given us less than three months to have this project constructed and operating. The site we found was close to roads, rail, and a port. It was an old shipyard. It was large, expansive, and still had some structures that could be incorporated into the design of the pig waste facility. There was a large building on-site that could accept and store waste. That could reduce odors and act as a central office. Also, it was zoned for industrial usage and had no homes in the vicinity.[28]

Next, we would need European Union (EU) and Belgian (and Flemish) permits: air, water, noise, soil, and odor. We spent weeks preparing permits and attending briefings with EU and Belgian regulators in hopes of obtaining bureaucratic approval. The air and water permits were granted without opposition. Odor would be a little trickier.

In the late 1990s, the EU had passed several "olfactory" ordinances to reduce the nuisance of odors.[29] The precise measurement of offensive odors can be difficult.[30] International efforts to compose and promulgate standards that were universally recognized had largely failed.[31] Think about how food from other cultures smell. Sometimes wonderful. Sometimes like mud. How could a government regulate and ensure compliance on something so relative as smell?

So, we hired an environmental company out of Frankfurt to conduct the analysis.

D. Reggie joked, "I hope this German is not some guy with a big nose, that takes a couple of sniffs, tells us to plant shrubs, collects his money, and leaves."

Yet that was what he was and did. We paid a German company almost $50,000 to conduct a "smell-survey" for the property. He told us to plant shrubs. He left and never came back.

Bills were starting to accumulate. How would we finance this? Blobby came through with the money. We had secured the contracts, the land, and now were on our way to getting all our permits. When we got back to the UK, Tryand felt it was time to start the work permit process for Juan and me. That meant negotiating a salary and benefits package. Streetman and Selby were completely against this. Why should the "experts" get paid? They were really against Juan and I having large shares of stock in the newly formed company. Why can they not use their hard work and sweat to barter for equity? Neither felt that our past efforts should be compensated.

I was disgusted.

"Well Danny," I queried, "What do you bring to the table? What is your role besides complaining all the time? Blowing your nose? Do have any expertise here?"

It was that day that I realized what Danny Streetman was: a loathsome bully.

Selby was a perspicacious and an astute businessman. His concerns were for the deal. Nothing personal. Just business.

"How about some combination of salary, benefits, and equity?" Ian offered.

We were allowed to discuss later the following day. Juan wanted to simply make a stand and play 'hard ball' in the negotiations. I felt that the process so far had been amicable, and we should not assume the worst. Nor should we accept whatever was offered. If we did that, then we would leave too much on the table.

"So, Juan, Johnny Baby, what have you decided?" George canvassed.

Against everything we had agreed to do, Juan burst in, "I cannot accept anything less than $5,000 per month."

I just sat there. Dumbfounded.

"And you?"

Everyone starred in anticipation.

"My time, experience, and value to this project are worth more than that."
Silence.

A dry hush like a whisper in the desert.

Juan was still thinking like a graduate student interviewing for his first job. Maybe he was overwhelmed by the National Liberal Club's leather chairs and marble floors. I was hardly a seasoned negotiator, but I had started companies, sold companies, and raised money based on a drawing from a cocktail napkin. I was incensed with Juan.

"I think we can do better than $5,000 per month." Even Streetman and Selby agreed with that.

We eventually agreed to a much larger number (see Figure 8.1 on the following page), with benefits, executive director positions, and a fair distribution of stock and other options.

I spoke with George later and he agreed that Juan needed to stick with engineering. Tryand started the work permit application us both. I would never have agreed to something where Juan got less than I did regardless of him trying to weasel me out just a few weeks earlier.

Back in New Orleans, Juan and I began the process of relocating. We settled on Belgium as the ideal location. Quality of life is wonderful in Belgium.[32] Also, this is where the pig facility would be located. It made no sense to "commute" from the UK to France to Belgium every day or even once a week. Juan wanted to move to Ostende. I picked Bruges. By train, Bruges is less than 15 minutes from Ostende.

Besides, Bruges is beautiful. The house I rented was built in the 1600s and was less than five minutes to the city center. It had three bedrooms and one bath. My deposit was $900, and my rent was $4500 for three months.[33] It featured a courtyard laced with greenery and some privacy. The reflections of the night sky shimmer across the water and can be seen from almost any window in the house. The city canals meander beside old cathedrals and outdoor markets. It ranks as a United Nations World Heritage Site.[34] There are

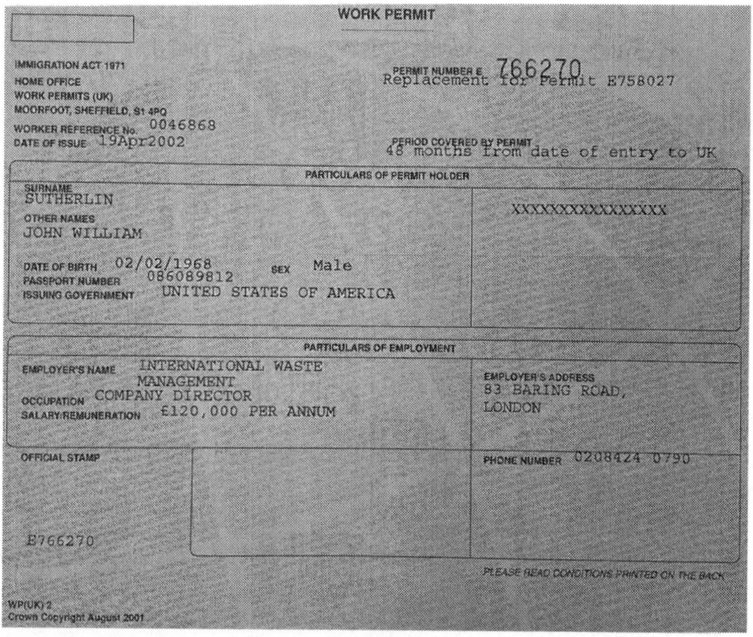

WORK PERMIT

IMMIGRATION ACT 1971
HOME OFFICE
WORK PERMITS (UK)
MOORFOOT, SHEFFIELD, S1 4PQ
WORKER REFERENCE No. 0046868
DATE OF ISSUE 19Apr2002

PERMIT NUMBER E 766270
Replacement for Permit E758027

PERIOD COVERED BY PERMIT
48 months from date of entry to UK

PARTICULARS OF PERMIT HOLDER

SURNAME
SUTHERLIN
OTHER NAMES
JOHN WILLIAM

DATE OF BIRTH 02/02/1968 SEX Male
PASSPORT NUMBER 086089812
ISSUING GOVERNMENT UNITED STATES OF AMERICA

XXXXXXXXXXXXXXXX

PARTICULARS OF EMPLOYMENT

EMPLOYER'S NAME INTERNATIONAL WASTE MANAGEMENT
OCCUPATION COMPANY DIRECTOR
SALARY/REMUNERATION £120,000 PER ANNUM

EMPLOYER'S ADDRESS
83 BARING ROAD,
LONDON

OFFICIAL STAMP

E766270

PHONE NUMBER 0208424 0790

PLEASE READ CONDITIONS PRINTED ON THE BACK

WP(UK) 2
Crown Copyright August 2001

Figure 8.1. The UK Work Permit, where we began seeing some financial benefits from the research. Courtesy of the Author.

colorful festivals all year round. International cuisine is available.[35] As the saying goes, "Everyone eats well in Belgium!"[36] And Brussels, the capital of the EU, is a little more than an hour and the cost is less than ten dollars at the time. Paris is less than three hours away from Bruges. London is only three and a half hours by Eurostar.

Moving to the EU is challenging for non-EU country citizens. Belgium required a work VISA, a criminal background check, and a sexually transmitted disease test. Because I was married with a child with another on the way, we started looking for schools and hospitals.

"Have you started getting your documents together?" I asked Juan. Since he was Ecuadorian, I knew that the process would be more tedious and longer. He also was married and had a dog. Getting the dog into Belgium was going to be arduous. Not impossible, just burdensome. It was more difficult to bring a dog into Europe than a child.

The first signs of reluctance showed on his face like a humiliated youngster. It was not anger. It was a sadness coupled with disappointment. His wife had been offered a job in California. He wanted to be supportive of her career. I agreed. But it meant he would have to spend time away from her in a foreign country half-way around the world. That is no way to raise a family.

There was still time, though.

In the meantime, we needed to get ready for our first presentation of the project before a proper funding agency.

Streetman and Selby wanted a backup to Blobby.

They were unquestionably correct.

George and Jim had set up a meeting with one of Europe's largest banks.

We were all headed to Frankfurt.

This was the presentation to secure more traditional financing through a proper financial institution and not with some "Jabba the Hut" with tales of Morgenthau notes and "black boxes" filled with Nazi gold or Treasury notes.

What utter rot!

Streetman and Selby were always cynical and mistrustful of Blobby. They had come up with that terrible nickname for him and his wife. They were horribly meanspirited.

That did not make them wrong.

Maybe they were right in being dismissive and skeptical.

But at least we had an alternative. A solid German bank with generations of experience and contacts throughout the world.

This could supplement or replace Blobby. We had an alternative.

Right?

Or did we?

We were heading to Frankfort to make the deal.

NOTES

1. The American Presidency Project. Executive Order 6102—requiring gold coin, gold bullion, and gold certificates to be delivered to the government. Located at https://www.presidency.ucsb.edu/documents/executive-order-6102-requiring-gold -coin-gold-bullion-and-gold-certificates-be-delivered. Accessed 22 June 2022.

2. International Churchill Society. Leading myths—Cape Town gold. Located at https://winstonchurchill.org/publications/finest-hour/finest-hour-173/leading-myths -cape-town-gold-a-churchill-myth-in-reverse/. Accessed 3 July 2022. Rumors about a greedy American government under FDR put forward the notion that the US was attempting to seize British territories in payment under the "lend-lease" program.

3. PBS. The American experience. Henry Morgenthau, Jr. (1891–1967). Located at https://www.pbs.org/wgbh/americanexperience/features/holocaust-morgenthau/. Accessed 22 June 2022.

4. Federal Reserve History. Henry Morgenthau. Located at https://www .federalreservehistory.org/people/henry-morgenthau-jr. Accessed 23 June 2022.

5. Ibid

6. Morgenthau Jr, H. (1944). Bretton Woods and international cooperation. *Foreign Affairs, 23*, 182.

7. Levy, H. (2010). *Henry Morgenthau, Jr.: The remarkable life of FDR's secretary of the Treasury.* Simon and Schuster.

8. New Orleans 27/7 since 1718. Located at https://www.neworleans.com/blog/ post/guide-to-bayou-saint-john/. Accessed 11 June 2022.

9. Fair Grounds Race and Slots. Located at https://www.fairgroundsracecourse .com/. Accessed 11 June 2022.

10. Degas Home & Museum. Located at https://www.degashouse.com/. Accessed 11 June 2022.

11. Louisiana Digital Library. Old Spanish custom house in New Orleans Louisiana circa 1930s. Located at https://louisianadigitallibrary.org/islandora/object/state-lwp %3A2104. Accessed 11 June 2022.

12. Goodell, J. W., & Vähämaa, S. (2013). US presidential elections and implied volatility: The role of political uncertainty. *Journal of Banking & Finance, 37*(3), 1108–1117.

13. DEFRA. Located at https://www.gov.uk/government/organisations/department -for-environment-food-rural-affairs. Accessed 27 June 2022.

14. National Liberal Club. Located at https://nlc.org.uk/. Accessed 4 March 2022.

15. Cramton, R. C. (1985). Preparation of lawyers in England and the United States: A comparative glimpse. *Nova LJ, 10*, 445.

16. The Independent. Guinness World Records: How the Irish brewer became an authority on firsts, feats, and pub trivia. September 5, 2018. Located at https:// www.independent.co.uk/arts-entertainment/books/news/guinness-world-records-new -edition-history-origins-brewer-ireland-a8523941.html. Accessed 23 June 2022.

17. Guinness World Records. Located at https://www.guinnessworldrecords .com/about-us/our-story. Accessed 23 June 2022. It should be noted, "Although the books never did tackle this original question—owing to their focus purely on

world records—the red-breasted merganser would be the most likely answer; it is fully migratory and still occasionally hunted."

18. Toole, J. K. (1987). *A confederacy of dunces.* 1980. New York: Grove Weidenfeld. Blobby never mentioned "theology or geometry" though.

19. The Guardian. Meet Ting Tong, new citizen of Little Britain. November 10, 2005. Located at https://www.theguardian.com/media/2005/nov/10/broadcasting .bbc1. Accessed 6 July 2022. It is hard to imagine how racist, and terrible it was to call Blobby's wife by the name of Ting Tong. Ting Tong was a mail-order Thai transexual that the main character bought. It is hard to believe that such a show would be on the air today. As with Blobby, I never called her Ting Tong.

20. 12 U.S. Code § 305—Class C directors; selection; "Federal reserve agent."

21. Research demonstrates aspects of his assertions. Meagher, R. F. (2013). An international redistribution of wealth and power: A study of the charter of economic rights and duties of states (No. 21). Elsevier. Bach, G. L., & Stephenson, J. B. (1974). Inflation and the redistribution of wealth. *The Review of Economics and Statistics,* 1–13. Beetsma, R., Cukierman, A., & Giuliodori, M. (2016). Political economy of redistribution in the United States in the Aftermath of World War II—evidence and theory. *American Economic Journal: Economic Policy, 8*(4), 1–40.

22. Dooley, M. P., Folkerts-Landau, D., & Garber, P. (2004). The revived Bretton Woods system. *International Journal of Finance & Economics, 9*(4), 307–313.

23. Morgenthau, H., & Balakian, P. (2003). *Ambassador Morgenthau's story.* Wayne State University Press.

24. Kochan, L. (1997). Jews for sale? Nazi-Jewish negotiations: 1933–1945. *The English Historical Review, 112*(448), 1022–1024.

25. James, H. (2001). *The Deutsche Bank and the Nazi economic war against the Jews: The expropriation of Jewish-owned property.* Cambridge University Press.

26. James, H. (2001). *The Deutsche Bank and the Nazi economic war against the Jews: The expropriation of Jewish-owned property.* Cambridge University Press. This should not have taken decades to resolve since documentation during this period existed. Dean, M. (1933). Robbing the Jews. The confiscation of Jewish Property in the Holocaust, 1945. However, there is still no accounting of the stolen books from this period. Glickman, M. (2016). *Stolen words: The Nazi plunder of Jewish books.* University of Nebraska Press.

27. Rednor, J. A. (1978). Who owns nonregistered bearer bonds? *The CPA Journal (pre-1986), 48*(000010), 62.

28. The remoteness of the site did have some disadvantages. Because no one was in proximity, nefarious activities often occurred here. We stumbled upon an amateur porn movie being filmed here once. It was not alluring or tempting. It was just repulsive.

29. McGinley, M. A., & McGinley, C. M. (2001). The new European olfactometry standard: Implementation, experience, and perspectives. In Air and Waste Management Association, Annual Conference Technical Program, Session No. EE-6b: Modelling, Analysis, and Management of Odours.

30. Van Harreveld, A. P., Heeres, P., & Harssema, H. (1999). A review of 20 years of standardization of odor concentration measurement by dynamic olfactometry in Europe. *Journal of the Air & Waste Management Association, 49*(6), 705–715.

31. Schulz, T. J., & Van Harreveld, A. P. (1996). International moves towards standardisation of odour measurement using olfactometry. *Water Science and Technology, 34*(3–4), 541–547. This remains a misleading pursuit: regulating odor.

32. Van den Bosch, K. (2019). Perceptions of the minimum standard of living in Belgium: Is there a consensus? In *Empirical Poverty Research in a Comparative Perspective* (pp. 135–166). Routledge.

33. I found the property listed on several travel destinations. It is still for rent. However, today the amount would be much higher. Instead of $50 per night, it is more than three times that. Also, the landlords have since updated the decor to a more contemporary look, especially the kitchen. I found the old renters contract while digging through memorabilia from this era. I then went online to look at some pictures.

34. Bruges. UNESCO Site. Located https://www.visitbruges.be/unesco-world -heritage. Accessed 23 June 2022. Bruges is often called the "Venice of the north." I think that does little justice to a city that is just fine being itself and not compared to anyone else. The festival seasons are diverse, and the colors are so vibrant. It is one of the most beautiful places that I have ever traveled and I often long to return.

35. Scholliers, P. (2008). Food culture in Belgium. ABC-CLIO.

36. Waterman, T. (2011). The flavour of the place: Eating and drinking in Pajottenland. In *The Landscape of Utopia* (pp. 40–56). Routledge.

Chapter 9

Embarrassment in Frankfurt

We were all excited. Bursting and overjoyed. Those were the collective emotions. I can remember discussing our upcoming meetings with exuberance. Our calendar showed nothing but opportunities for triumph. A quick flight from London to Frankfurt allowed us to brief ourselves and prepare for an invaluable meeting. We had studies, reports, slides, handouts . . . This could change everything for the project. Could this be an alternative to Blobby? Maybe to supplement Blobby?

We were about to find out.

The Bundesbank in Frankfurt is one of the largest in Europe.[1] With more than $250 billion, it dwarfs many older banks in the European Union (EU). German banks in the early 2000s were still a central part of the financial structure of their economic system.[2] For foreign nationals looking for EU financing, German banks can direct a client through the maze of regulations and laws and provide management expertise to ensure that a project is successful.[3] Many talk about the unique nature of German banks and how they are less likely to experience the "boom-bust" cycles of capricious investing.[4] Yet the "German uniqueness" also makes it so risk averse that it was the least likely bank to fund a pig poop project.[5]

We had no idea of this. We should have known, though.

The meeting commenced with the generic formalities of any meeting with names, country of origin, level of expertise, and our role in the project. George Miller took the lead. He spoke flawless German. His wife was German. His children spoke German. For the sake of everyone else, the investment officers agreed to speak English. We were grateful. They seemed genuinely ambitious to work with us.

It was a good sign.

But it was the last good thing that happened that day.

Our inaugural comments seemed to fall flat. Our enthusiasm had not translated well into German. No one was moved. They were grossly underwhelmed and inattentive. It was fatigue like when you hear the same tired joke for the 100th time. Their weariness was evident and palpable.

"Yes, thank-you. We are aware of the problems of too many pigs in too small of a farm. In Germany we have many universities and farms working on this problem. In the US, you Americans are also working on this problem."[6]

Not a great start.

They did not need Americans or Brits or anyone else explaining this problem to them. They understood the issue and wanted us to get to the point. Fair enough. No one wants to be told how to tie his own shoes in his own house. We moved on to more technical and practical matters where we could establish our competency.

Our ensuing statements did not seem to have the desired impact. We discussed the research, technology, patents, and how we could scale-up from a demonstration unit on the back of a trailer to a system at the Port of Ostende. We discussed our meetings at the EU with agricultural experts in Brussels. We then handed out brochures and showed copies of studies that had 'proven' that our project was cost-beneficial. And could be profitable.

They were not impressed.

The committee, which was three or four men in black suits with slick bright blonde hair and piercing blue eyes and a dark-haired secretary taking notes, who sat stone-faced in the corner. I assumed that was the lawyer. He never smiled. It was that feeling you get when you tell a joke at a party that no one gets the punchline. And the more you explain it, the less amused and the more uncomfortable everyone appears.

Someone coughed. It was that dismissive and contemptuous cough that an older Aunt does during dinner when someone says something foolish or inappropriate. The kind of cough that means, "Let's just move on. And, please stop talking."

Mortification.

If I could have melted into the furniture, I would have.

That is the best I can explain. Beyond humiliation. Complete embarrassment. I can remember wishing I was back in the third grade playing kickball with Sister Sophia. Or even in detention picking up trash after school. Anywhere but here. I was a child among men. I did not belong here with my stupid idea. My idiotic business plan revealed how little I knew. I needed to go back to New Orleans and stick to something I understood. Maybe I should teach High School Civics or debate contemporary political issues with students.

These were serious people.

It was not even that the Bundesbank officials were uninterested. They would have needed to feel something towards us to rise to the level of disgust. It was disdainful, malicious contempt. We had wasted their morning and now we were going to pay the price. Still, we slouched on to the next part of the presentation.

We meandered through our juvenile, cartoon-like PowerPoint to the financials. Here, we "showed" double-digit returns on their investment and how we could "guarantee" long-term profits with "no risk." For real? Did we really say that? Yep. In Voltaire's Candide, there is a hopelessly optimistic character that has given us a delightful expression: "Panglossian."[7] Foolish optimism lacking in rationality or reason. That was us. Don Quixotes of crap!

How had we got to this point? Was it pride? Vanity? We had sat around listening to our own voices without ever considering the alternative. Prior to this meeting, we should have done a focus group with comparable experts and reacted to their questions. But here we were.

A couple of them scribbled on their notepads. Probably nothing related to what they had just saw.

Then they got nasty.

The rapid fire questioning destroyed me.

"Dr. Sutherlin, what makes you think that any project is risk-free?"

"Who can guarantee such high rates of return?"

"How could we justify giving you a standard business loan much less an investment?"

"Why would anyone risk money on such a project?"

"What made you think this was a good idea?"

"Why would you wear those shoes with that suit?"

The last comment was a little off the mark and too personal. It did not matter. My ideas and shoes were all wrong. The Germans were lucid and clear. The Bundesbank had no recommendations and no interest in this project.

Not a thank you for considering us.

Not a good luck with you project.

Not a can we validate your parking.

Nothing.

The meeting unostentatiously ended.

I heard one of them mumble, "Dummkopf," as he left the conference room.

I was so glad to get out of there. My dissertation defense had been a cinch in comparison to this nightmare. Our hallucinations fed our delusions and had caused so much destruction. The project seemed to be annihilated. We must have breathed too much ammonia at the pig farm, and it crippled our higher reasoning skills. This had been an idiotic venture with no silver lining. The sky was no longer as blue. The sun shined a little less brightly. I felt shorter. I even needed new shoes.

I do not remember any of us speaking until we got back to London. We limped to our respective places. Naturally, it was raining. The rain felt colder and harder somehow. I am glad it rained. No one can see you crying in the rain. I was too hurt to cry. My surgery had ripped a chunk of my intestines out. This was worse.

Finally, George spoke, "Gentlemen, I am glad we got that one over. In any deal there are meetings like this one. Terrible setbacks with nothing to show for our efforts. At least we got this one over sooner than later. This was bound to happen at some point."

I cannot believe he had the strength to say that.

I could barely breathe.

I did not have the inner strength to utter a syllable. We all needed a pep talk. George sensed it.

"Johnny baby, there is a difference between a blunder and a fiasco."

"What is it?"

"I don't know."

I looked up.

George's bright eyes and enormous smile allowed me to see inside his even bigger heart. He had been knocked around but not defeated. This was a temporary condition. He could make light of the situation. Maybe there was still hope.

I could breathe a little easier.

"What do we do, Georgie-baby?" I decided to take a page from him.

"We dust off our shoulders and get ready for our next presentation."

He, of course, was right. We had to quickly change our presentation and attitudes.

Next door to Belgium is the Netherlands. In Holland there is a collection of smaller banks that understand and invest in agriculture projects. Belgium banks were never considered, and everyone agreed they were neither big enough nor able to manage such a complex project.[8] I had no idea and could offer no assistance here. This was "the boys from Europe's" field of proficiency.

Danny Streetman (the conservative politician and farmer) knew people at Rabo Bank.[9] Dutch banks function in a similar role of aiding prospective clients with management training, but Rabo Bank also served in as an extension service provider with area farmers.[10] According to research, "Extension is a communication activity that aims to stimulate a specific audience to acquire relevant knowledge about specific issues."[11] Dutch banks also promote best new technologies and implementation of sustainable agriculture.[12] We were, hopefully, doing both. Perhaps the most important issue was that the Dutch were aggressively seeking solutions to the pig waste problem.[13]

Our meeting was set.

We would travel to Utrecht. Just south of Amsterdam, Rabo Bank was about three hours from Ostende. This time we spent some time along the Flemish coast before driving into the Netherlands. The coast is about 42 miles long and has shops, bars, and restaurants that feature locally grown foods and locally brewed beers. I even found another one of those chocolate shops with the seashell treats.

The presentation was almost identical to the disastrous one in Frankfort. There would be less cartoons and more pictures from actual farms and the proposed sites in Belgium. We dropped the handouts. The reception was different. Rabo Bank got it. They understood right away. They were aware of the pig waste problem, knew it needed to be solved, and were actively looking for projects to fund. The Netherlands was working with France and Denmark on these issues.[14] They expressed gratitude for our time. They shook our hands. They validated our parking. No one commented on my shoes.

What just happened?

Was I hallucinating?

How could we get such different responses using the same presentation?

Not every project is a good fit for every group of investors. When selling technology or environmental opportunities to investors, every group must be evaluated as an independent entity.[15] All banks are not the same. All environmental investors are not the same. Everyone has a different tolerance for risk, different motivations, and expectations. Rabo Bank wanted to take a chance on solving this international water crisis. They had a social responsibility that drove their investment portfolio.[16] It was not just profits. They wanted to make a difference.

They thanked us for our hard work.

We pledged our solidarity in working together.

"Let's find a way forward."

Everyone agreed.

The trip back the UK was much better this time.

We split up so that Streetman could go back to south England, and we could travel back to South London (Bromley).

There was a feeling of exhilaration and euphoria among the group. George's eyes beamed a bit brighter. His teeth flashed whiter. We were fully recovered from the embarrassment in Frankfurt. I felt like a human being again. My embarrassment had evaporated.

"Johnny baby, what did I tell you?"

He was right.

The thrashing in Frankfurt was just an unpleasant, almost distant memory. Had it only been a couple of days? Like being food poisoned during Thanksgiving. The immeasurable pain and discomfort. But now it was over. We survived. Today was Christmas. Time to open our presents. Have a nice meal. Kiss our friends and family. Toast to the New Year.

Cheers!

Things had changed so fast.

Fortune had smiled on us.

The fates can be capricious and fickle.

Opportunities are never permanent.

Blobby resurfaced. Orca had come up for air. He had great news. Prior to our Frankfurt debacle, Blobby had expressed "procedural delays" from the Federal Reserve. We had received some monies (about $500,000) but not the full amount (more than $5 million) needed to move forward. According to Blobby, all the delays in accessing the funds had been cleared. George was needed to go to New York and finalize the details for the first project in Ostende. There was a catch. Bristow wanted us to build a similar facility in Brazil as well. He knew this was off the "beaten pig trail," but the Federal Reserve "demanded it."

None of us were excited. Streetman and Selby especially. Here we had rescued the project from the depths of financial Hell only to have Blobby drag us back down. George briefed the team and was headed to New York in a couple of weeks.

"We need to keep as many options open until we get full funding," he pleaded.

This was true. Until we had $5 million to fund the full project, there was no point in refusing to meet with another potential investor. But Blobby?

We went back to America to await the news of the meetings.

Days went by and there was no news.

Nothing.

Nicholas Bristow never showed up.

He did not answer his phone.

His wife did not have any idea where he was. She was in Montreal on another project.

She was worried.

George was perturbed and troubled.

More so than when we were given a proper throttling by Bundesbank.

Where was Blobby? His health tormented our thoughts. His bulging body was a ticking time bomb. Maybe he died? A heart attack or a stroke was undeniably possible. Maybe even likely considering everything.

I would not interpret his actions and comments correctly until years later. At the time, George sounded and looked worse. He sounded tired. Beaten down. His voice was like gravel and sand. Our telephone conversation was distressing.

Where was Bristow?

Then the news broke. Blobby had taken ill following the Federal Reserve meeting in Washington, DC. The stressful meeting had caused him to blackout. He was in a hospital in Baltimore. He had been unconscious for days. His wife called George within a few days after he traveled back to London.

George announced, "We found Blobby."

Had the fates shown us a path forward? With Blobby sick, what else should we do except pursue the plan put forward by Rabo Bank? Besides, Blobby's illness seemed suspicious to everyone. Rabo Bank was real, though. They had conditions as any lender would. They suggested a package of loans and equity financing. They had an investment group ready to participate. Ian and Danny had always supported conventional financing and not Blobby.

They were right.

I had lost faith in Blobby after this.

Something just did not feel right. He vanished and then showed up in a hospital. Maybe. Yet, his wife stayed in Montreal while her husband laid in bed possibly dying. Doubtful. You never want to get inside someone's marriage. I had never met Bristow's wife. Who knows?

I wanted real answers.

I was about to uproot my family and move to Bruges.

I needed more than Blobby's word.

I was asking tough questions.

Jim Tryand stepped up. He expressed concerns about Blobby as well. He would contact friends of his at the Federal Reserve and the World Bank. He would get answers. We would not take another step with Blobby without hard evidence and more money. Tryand's due diligence would answer many questions. I believed in Tryand and relied on his expertise and contacts to get the answers all of us were demanding.

Blobby had financed our project so far. But we needed much bigger money to take the next step. The delays from the Federal Reserve, his health, and now a Brazilian project sounded more suspicious every day. It may seem obvious now but faced with the prospects of building 10 systems around the world, or a smaller one in northern Belgium, it is easy to delude yourself thinking about all of the money you could make. Greed is a terrible sin.

The project was a crossroads.

Which route would we take?

An "angel" investor with access to billions of dollars?

Or a Dutch bank ready to finance one system?

We saw what we wanted to see.

We also were losing sight of why we had traveled down this road in the first place.

Was this still an environmental project?

Or just another business deal?

NOTES

1. Bundesbank. Located at https://www.bundesbank.de/en. Accessed 12 March 2022.

2. Hackethal, A. (2003). German banks and banking structure (No. 106). *Working Paper Series: Finance & Accounting.*

3. Deeg, R. (1998). What makes German banks different. *Small Business Economics, 10*(2), 93–101.

4. Behr, P., & Schmidt, R. H. (2016). The German banking system. In *The Palgrave handbook of European banking* (pp. 541–566). Palgrave Macmillan, London. Hackethal, A. (2003). German banks and banking structure (No. 106). *Working Paper Series: Finance & Accounting.* Hüfner, F. (2010). The German banking system: lessons from the financial crisis.

5. Vitols, S. (1998). Are German banks different? *Small Business Economics, 10*(2), 79–91. Hoggarth, G., Milne, A., & Wood, G. E. (2001). Alternative routes to banking stability: A comparison of UK and German banking systems. In *Financial competition, risk and accountability* (pp. 11–32). Palgrave Macmillan, London. Fischer, K. H., & Pfeil, C. (2003). Regulation and competition in German banking: An assessment (No. 2003/19). CFS Working Paper.

6. Hansen, B., Alrøe, H. F., & Kristensen, E. S. (2001). Approaches to assess the environmental impact of organic farming with particular regard to Denmark. *Agriculture, Ecosystems & Environment, 83*(1–2), 11–26. German universities and government funding support research throughout the region. Germans were actually taking the research to places that we had not considered. Hamscher, G., Pawelzick, H. T., Sczesny, S., Nau, H., & Hartung, J. (2003). Antibiotics in dust originating from a pig-fattening farm: A new source of health hazard for farmers? *Environmental Health Perspectives, 111*(13), 1590–1594. We assumed that CAFOs could be modified or lagoons treated. They were rethinking the entire sector. We gathered lots of data from European studies. Some were only marginally related to our efforts but presented interesting findings. Halberg, N., Verschuur, G., & Goodlass, G. (2005). Farm level environmental indicators; are they useful? An overview of green accounting systems for European farms. *Agriculture, Ecosystems & Environment, 105*(1–2), 195–212.

7. Scherr, A. (2006). Candide's Pangloss: Voltaire's tragicomic hero. *Romance Notes, 47*(1), 87–96.

8. Pintjens, S. (1994). The internationalisation of the Belgian banking sector: A comparison with the Netherlands. In *The competitiveness of financial institutions and centres in Europe* (pp. 301–311). Springer, Dordrecht.

9. Rabo Bank. Located at https://www.rabobank.com/en/home/index.html. Accessed 4 March 2022.

10. Wielinga, E. (2000). Rural extension in vital networks, changing roles of extension in Dutch agriculture. *Journal of International Agricultural and Extension Education, 7*(1), 23–36.

11. Ibid.

12. Röling, N.G. (1988). *Extension science: Information systems in agricultural development.* Cambridge: Cambridge University Press.

13. Boinon, J. P., & Hoetjes, B. J. S. (1999). Netherlands: From waste quotas to pig quotas. *L'agriculture européenne et les droits à produire.*, 271–287.

14. Poulsen, H. D., Jongbloed, A. W., Latimier, P., & Fernandez, J. A. (1999). Phosphorus consumption, utilisation and losses in pig production in France, The Netherlands and Denmark. *Livestock Production Science, 58*(3), 251–259.

15. Mason, C. M., & Harrison, R. T. (2003). "Auditioning for money": What do technology investors look for at the initial screening stage? *The Journal of Private Equity, 6*(2), 29–42.

16. Jeucken, M. (2010). *Sustainable finance and banking: The financial sector and the future of the planet.* Routledge. A more recent work includes the following. Bouma, J. J., Jeucken, M., & Klinkers, L. (Eds.). (2017). *Sustainable banking: The greening of finance.* Routledge.

Chapter 10

The Vatican and a Night of Jazz

Voodoo is the common thread throughout this chapter. Maybe the project was jinxed. Not with the soft, tourist Voodoo in some cheap shop that sells mojo-bags, incense, and t-shirts with Marie Laveaux stamped in pastel. The real Hoodoo. All the forces of weird were about to hit us. Unhinged and irrational all together.

George Miller and Jim Tryand insisted on coming to New Orleans to meet with Juan and me. They proposed to discuss the latest Nicholas Bristow scheme. Blobby's lies were starting to weigh me down. It had taken too long, but I was wising up to his hustles. The oppression was like a merciless, and relentless dark cloud hovering over the project. Always threatening to rain. Teasing us about the sun. A trickle of money flowed, but we needed large-scale funding. I have learned that when a funding agent defaults once, you should cut your losses and move on quickly. They either have the money or they do not. They are either prepared to invest or they are not. No games.[1]

I was unsure of the purpose of the meeting, but I enjoyed seeing George and Jim.

So, I scheduled their arrival to coincide with a Russian feast at the Plimsoll Club (then) at the top of the World Trade Center in New Orleans.[2] Also, I thought that a night of Jazz at the best venue in New Orleans would be a great cure for what would be a long week of meetings. And, since Astral Project was playing, what could be better? George had opened his home to me and my wife and showed extraordinary hospitality. I intended to extend all amenability and receptiveness to him.

George and Jim came to my house and met my family. All were enchanted. Having them come to us in America was a gracious gesture on their part. But we were not meeting for social reasons. George enjoyed the Russian food, vodka, and music at the Plimsoll Club. My wife had joined us for an afternoon and was impressed with George.

The Jazz performance was, as expected, a success. Everyone sat stunned to see live Jazz in the City that invented Jazz.[3] We shambled through the French

Quarter looking for the ghosts of Jelly Roll Morton and Louis Armstrong. Everything we did socially was pleasurable.

Our meetings were a little different.

There was illumination but no clarity.

George and Jim still seemed locked into the Blobby funding mechanism. Their fixation would be made manifest later.

I braced myself.

"Johnny baby, Blobby has stepped up in a way that exceeds all of our dreams," he finally divulged.

"Tell me," I said nonchalantly. Not bored, just guarded. I had heard so much hog wash before this week. I was jaded and my interest in Blobby had faded.

"Jim, you had better explain," George relinquished the floor and yielded time to the "right honorable gentleman from Michigan."

Jim Tryand wove a story that was astonishing, often spiked with truth, but always deliberately deceptive for the purpose of keeping Blobby in the picture. "Back before the collapse of the Soviet Union, there were refineries dedicated to providing fuel for the Russia Navy. These were strategically located in Portugal and Brazil. Now, there is no Soviet Union. But the oil was stored there for years, almost a decade. Slowly, world demand has drained these reserves and left behind abandoned refineries."[4]

"Ok, I follow." I understood the environmental horrors of the Soviet system as I had conducted my master's thesis on Russian ecocide in the Caucasus region.[5]

Jim continued, "These refineries are environmental pits and must be cleaned up and reused for something. There are limits because of the existing pollution and health matters. That is where we come in."

"Go ahead."

"The Federal Reserve wants to kill many birds with one stone. Cleanup these refineries, build sustainable businesses, and support our projects. This could produce a lot of jobs and those countries would be grateful to the US."[6]

"Why us?"

"Blobby."

"Blobby?"

"Blobby, is an agent with the Federal Reserve. I checked him out. I have verified him. He is in-charge of projects. He has been pitching our project for months. And there is another piece."

George was watching me to judge my expression.

"Yeah, what?"

"The Vatican."

"What?" I sat stone-faced as disbelief and disappointment washed over my mind.

For centuries, the finances of the Vatican have been shrouded in mystery.[7] Besides donations from church goers, a big piece on the Pontiff pie comes from investments.[8] Some scholars have made wild claims over the years, but the consensus is that the Vatican, the world's smallest state, has per capita wealth beyond any Middle Eastern Sheikh.[9] Like other micro-states, say, Luxembourg or Malta, there was always so many myths floating around.[10]

Thus, speculation about Vatican investments ranges from the absurd to the barely plausible. One scholar even linked Papal banking policy to the Great Depression.[11] Conspiracy theories run amok when it comes to the Church. More contemporary research suggests that the Church profited from being too cozy with the Nazis.[12] It seems there is no end of throwing stones at the Catholic Church. Notwithstanding prejudice against Catholic charities, it remains one of the most critical non-governmental organizations (NGOs) in the world that feeds millions.[13]

Putting aside rumors and inuendoes for the moment, what did the Vatican have to do with us?

Jim Tryand rubbed his eyes, wiped his mouth, and continued, "They are in the same position as the Federal Reserve. Remember, it was the Italians that shot down the plane carrying the Morgenthau notes. The Vatican has been sitting on these notes. They wanted to blackmail the US to prevent any reparations for their complacency during World War Two. But all that has gone out the window. Now, they have billions of dollars in bearer bonds, and they want their pound of flesh."[14]

I thought through this.

My head hurt.

We were back to that hustle?

I could have chewed barbed wire and spit nails.

Not those damn Morgenthau black boxes again.

A pork Ponzi scheme?[15]

Was there a depth that was too low for Blobby?

Sensing my disbelief and frustration, Jim decided to provide the "actual proof." In his briefcase, he had signed document from His Holiness authorizing the Vatican Investment Committee (a real entity but not until years later)[16] to work with us. Gold stamps and notarized letters. The name on the documents was of a prominent European banker that many years later would be named to head the actual organization. Blobby had done his homework. The Vatican had allegedly extended to our project an additional $200 million subject to us building our next projects in Portugal and Brazil. Belgium is still the first project. The other two are for political purposes and it does not matter if they ever break even.

This was humanitarian investing.

The first "tranche" of money would be released within 90 days. In retrospect, I have noticed that international scam artists have their own coded language. Whenever I hear the word "tranche" from someone, I know they have either been scammed, or are trying to scam. Either way, I walk away. I did not know any better then. I was still an optimistic, naïve person.

Another word that puts me on guard is "exclusivity."[17] This is how people lock up your intellectual property (IP) for a period until they sell it, borrow against it, or raise equity investment monies. The other one is the phrase "first right of refusal."[18] If you are really wanting to participate, then just buy it. This type of chicanery is just to liberate a fool from his money using the best tool: jargon.[19] Not the ham-fisted Nigerian scam lingo about lost money and a dead General's daughter on life-support; this was sophisticated European-level scammers.[20]

Unfortunately, Americans make easy targets for financial con-artists. We are 40 times more likely to get scammed than have our cars stolen.[21] I have always wanted to believe it is because most Americans want to help others. That is certainly the case for many people. For others, it is just plain old greed. I wish that I had had more experience in dealing with these types of people. Years later, I read a praiseworthy work about how to spot a fraudulent deal.[22]

Blobby checked every box.

He was an unparalleled rip-off rat.

A hustler without a conscious.

Yet George was a proven international consultant with legitimate governmental and business experience. He was a cosmopolitan man, with financial competence and knew what he was doing. He had represented some of the world's largest corporations in waters so treacherous. He had swum with the sharks and lived to make another deal. Certainly his "red flags" were flying high.

Right?

Still the subterfuge now was the Vatican? A Papal con?[23] The Federal Reserve was not enough? Morgenthau and Nazi gold were *fata morganas* of the first order. Now we were getting a dose of religion and government? The chimerical nature of these latest editions of trickery were too much. I was about to explode. We were preparing to relocate our family to Belgium within a few weeks. I had passed the criminal background check, the sexually transmitted disease check, and the financial check. Juan had bought a storage container and was loading it to ship to the Port of Ostende. He had gotten a permit for his dog. He was days away from leaving New Orleans bound for the Flemish coast of Belgium.

Jim must have sensed my indignation,

I was about to walk away from George and Jim.

"Johnnie baby, there is more."

There was.

Jim provided it, "And, Blobby has deposited $250,000 in our company account so that we can complete additional analysis, studies, permitting. And so that you and Juan can started getting paid. Right away. Blobby knows that there have been too many delays and he has made it possible for us to begin work now."

I was stunned.

I had no answer for this.

It was completely unexpected.

Appreciated, but unforeseen.

Was it remotely possible that Blobby was real?

Morgenthau black boxes and Vatican gold?

There was no way this was real.

Still, the $250,000 was real.

And if Blobby was scamming someone, it was not me. It was not Juan. It could not be "the boys from Europe" as money was flowing into their account.[24] What kind of a scam pays money into your account? I had no answers. None of us had lost any money. Perhaps Blobby was using this project to rip-off others that we did not know. That was possible. Were we complicit in his crimes?

Then, George chimed in, "We need you sooner than later. There may be a delay in moving you all to Belgium permanently, but the work must start now. We, rather you, must complete the second patent."

I agreed. We had a device (or apparatus) patent. Now we needed to protect our IP on the process. Patents are only protections against infringement. We needed to lock up the IP by protecting the device ("Jurassic Park") and the "how" or the process.[25]

The money would come in tranches (that word again) until we broke ground in Ostende. We needed to secure all the pig waste, sewage sludge, and find markets for the fuel pellets we would be making.

Juan would be very happy.

This would be the proof he needed to get back fully committed to the project. He had lost faith and this would reengage him.

Nope.

Juan was miserably disenchanted.

His wife was done with "the boys from Europe." Blobby made her want to vomit. She did not want to hear George's name again. As such, she was going to take the job in California and Juan had no options left. He would have to go. He would look for an engineering job for himself. I never argued with him on this point. Of course, his wife was right.

Something else was weird, though.

Juan would move to California and remain involved in the project. It would take more of an effort, but he assured me that he was still 100% steadfast. Inconceivably, though, Juan refused to accept any compensation from the project. He would get reimbursed for travel related expenses, but never take a penny for his time.

"Why?" I asked.

"I do not want to get paid until the project is fully funded."

Maybe.

There had to be something else. I had no idea what it could be. I did not feel like arguing and found there was little point once he had made-up his mind. We had received a salary during the research phase of the project. So, why not now? Did Juan suspect that Blobby was a fraudster and not want to be too connected?

"Fine. How do you want to handle the next patent?"[26]

"I'll be in California. Once I am settled, then come out and you and I will write it."

I guess that is what I would have to do.

I would head to California in about a month or two.

In the meantime, I would go back to the UK.

George, Jim, and I needed to finish the pig poop contracts and get the facility permitted.

We had $250,000 in the bank with more coming.

Things were moving ahead.

Were they?

NOTES

1. Deevy, M., Lucich, S., & Beals, M. (2012). *Scams, schemes & swindles*. Financial Fraud Research Center. Another good article is as follows. Chariri, A., Sektiyani, W., Nurlina, N., & Wulandari, R. W. (2018). Individual characteristics, financial literacy and ability in detecting investment scams. *Jurnal Akuntansi dan Auditing, 15*(1), 91–114.

2. Quillen, Kimberly. The Times-Picayune. January 8, 2010. World Trade Center will move its Plimsoll Club to the Westin Hotel in March. Located at https://www.nola.com/news/business/article_b891d0ec-f0e9-54d3-bcd0-2b8df83e6d75.html. Accessed 8 June 2022.

3. Lomax, A. (2001). *Mister Jelly Roll: The fortunes of Jelly Roll Morton, New Orleans creole and "inventor of jazz."* University of California Press.

4. Dasgupta, B. (1975). Soviet oil and the third world. *World Development, 3*(5), 345–360. Here, the author show "the role played by the 'Soviet oil offensive' in the 1960s and its success in weakening the grip of the oil trade in the disintegration of the world-parity-pricing system, and the emergence of OPEC as a powerful factor in

the political economy of world oil. In the final section, the paper assesses the role of Soviet oil exports in the present world context and in the future." Again, there was an element of truth to the story by Tryand. And, as far as environmental concerns, Soviet policy towards their "allies" was an ecologically nightmare. Radelyuk, I., Tussupova, K., & Zhapargazinova, K. (2019). Assessment of groundwater safety around contaminated water storage sites. In *11th world congress on water resources and environment: Managing water resources for a sustainable future-EWRA 2019*. Proceedings.

5. Dumont, H. (1995). Ecocide in the Caspian Sea. *Nature, 377*(6551), 673–674. Krogh, P. F. (1994). *Ecocide: A Soviet legacy*. Zeisler-Vralsted, D. (2019). *Eurasian environments: Nature and ecology in Imperial Russian and Soviet history.* Edited by Nicholas B. Breyfogle.

6. Braunstein, S., & Lavizzo-Mourey, R. (2011). How the health and community development sectors are combining forces to improve health and well-being. *Health Affairs, 30*(11), 2042–2051. Again, a slight sliver of truth doled out to a hungry researcher.

7. Pollard, J. F. (2005). *Money and the rise of the modern papacy: Financing the Vatican, 1850–1950*. Cambridge University Press.

8. Lewin, E. A. (1983). The finances of the Vatican. *Journal of Contemporary History, 18*(2), 185–204.

9. Posner, G. (2015). *God's bankers: A history of money and power at the Vatican*. Simon and Schuster.

10. Eccardt, T. M. (2005). *Secrets of the seven smallest states of Europe: Andorra, Liechtenstein, Luxembourg, Malta, Monaco, San Marino, and Vatican City*. Hippocrene Books.

11. Pollard, J. F. (1999). The Vatican and the Wall Street crash: Bernardino Nogara and papal finances in the early 1930s. *The Historical Journal, 42*(4), 1077–1091.

12. McGoldrick, P. M. (2012). New Perspectives on Pius XII and Vatican financial transactions during the Second World War. *The Historical Journal, 55*(4), 1029–1048.

13. Barnett, M., & Weiss, T. G. (Eds.). (2008). *Humanitarianism in question: Politics, power, ethics*. Cornell University Press. During the French and then American fighting in Vietnam, for example, it was Catholic charities that stepped up to shelter locals. Pergande, D. T. (1999). *Private voluntary aid in Vietnam: The humanitarian politics of Catholic Relief Services and CARE, 1954—1965*. University of Kentucky.

14. An interesting quote at this juncture. The Merchant of Venice.

15. Frankel, T. (2012). *The Ponzi scheme puzzle: A history and analysis of con artists and victims*. Oxford University Press.

16. La Croix International. Vatican sets up Investment Committee to ensure ethical use of funds. June 8, 2022. Located at https://international.la-croix.com/news/religion/vatican-sets-up-investment-committee-to-ensure-ethical-use-of-funds/16207. 23 June 2022.

17. McHugh, C. (2015). The illusion of exclusivity. *European Journal of Philosophy, 23*(4), 1117–1136. There is a language unique to hustlers and con artists. It sounds like the hidden vernacular of international finance, but it is really used to limit clarity. Years later, Netflix would produce a movie called *The Laundromat* (2019).

This is close to the types of people I encountered. Once you shake hands with these types it is impossible to forget their style.

18. Walker, D. I. (1999). Rethinking rights of first refusal. *Stanford Journal of Business Law and Finance, 5*, 1.

19. Shadel, D. (2012). *Outsmarting the scam artists: How to protect yourself from the most clever cons*. John Wiley & Sons.

20. Schaffer, D. (2012). The language of scam spams: Linguistic features of "Nigerian fraud" e-mails. *ETC: A Review of General Semantics*, 157–179.

21. Munton, J., & McLeod, J. (2011). *The con: How scams work, why you're vulnerable, and how to protect yourself*. Rowman & Littlefield Publishers.

22. Fisher, K. L., & Hoffmans, L. W. (2009). *How to smell a rat: The five signs of financial fraud*. John Wiley & Sons.

23. There have been substantiated financial scandals involving the Vatican. Nuzzi, G. (2014). *Ratzinger was afraid: The secret documents, the money and the scandals that overwhelmed the Pope*. Adagio eBook.

24. Abagnale, F. W. (2002). *The art of the steal: How to protect yourself and your business from fraud—America's #1 crime*. Currency.

25. Lerner, J. (1994). The importance of patent scope: An empirical analysis. *The RAND Journal of Economics*, 319–333.

26. There was always discussion about whether to patent or not. As an environmentalist, I felt that protecting the IP of a product that could improve air, water, soil, and human life was important, but violated the essence of why we were doing the work. I was not and still am not against making money. I just felt that we should make this available to the industry without charge. Kieff, F. S. (2002). Patents for environmentalists. *Washington University Journal of Law and Policy, 9*, 307. Either way, there are mixed results for "green patents" in the marketplace. Environmentalists do not always make the best capitalists. Dong, S., Gong, H., & Liu, T. (2022). Environmental technology spillovers and green start-up emergence: The moderating role of patent commercialization policy and patent enforcement. *Environmental Science and Pollution Research*, 1–14.

Chapter 11

California Dreaming and
Another Piggy Patent

I had traded my overhauls and rubber boots for pinstripes and wingtips. The Swine Unit research seemed like a distant dream by now. I could still recall the pungent aroma of pig waste. Now, it was more like hog wash. How long it seemed since I had worked on the farm. Had it been years?

By the time I returned from the UK, Juan had relocated to California. Also, another disaster had befallen us all: 9/11. The tremors of September 11th would haunt, *inter alia*, the financial world for decades.[1] America had a new President. One that was less interested in the environment than reestablishing American hegemony.[2] Now, following the collapse of the Twin Towers, finding eco-friendly investors would be unlikely.[3] The world was in survival mode.[4] My reception in the UK was heartwarming and genuine. Even Danny Streetman shook my hand with sincerity and a sense of camaraderie.

I was given strict orders: complete the next patent.

So, I headed to California to see Juan.

He was loving the West Coast. Juan surfed. Perhaps that was as close to a religion as he had. He lived near San Juan Capistrano in a little Bohemian house that was comfortable for one person and cramped for two. But he was happy. Now, I was there. On our first day, we travelled south to San Onofre Beach. He surfed, and I sat in the sand having a beer. San Onofre is a clothing optional beach.[5] Nude beaches are not quite what people think. Most people look like they are waiting for the bus back to Middle Earth. All hairy Hobbits and chunky Trolls, no hot little Elves.

It took a full day to get back to the mindset of pig waste management. Hours and hours were spent reviewing all the literature that had once been instantly accessible. After all, we had become "big-time international deal-makers." We had the Vatican, the Federal Reserve, and Blobby backing us up. We were no longer covered in pig waste. The smell of money permeated the air now. Besides, we had been doing this for years. What could go wrong?

For one, lots of research had stepped into the vacuum following Hurricane Floyd. Lots of universities around the US and world had accessed US Department of Agriculture (USDA) grants or state appropriations (Iowa and North Carolina). Many patents had been filed since we got ours.[6] Novel approaches that we had neither knowledge of nor had experience to evaluate.[7] Universities studied lagoons and nutrients providing sustenance for plants and wetlands.[8] Unique applications with peer-reviewed assessments. We had Blobby.

Were we just dreaming in California?[9]

Had we missed the opportunity?

Maybe.

It did not matter, though.

We still had a job to do. We had made a pledge.

The second patent would not require weeks of research, countless edits, or too much coffee. It would flow better. But lagoons also flow and they still stink. The process was painful. We worked in a half-sized room off the kitchen and had little privacy. It was impossible to work efficiently while Juan's wife was home. So, we worked late at night and while she was at work. Our efforts resulted in another patent. This was filed and granted quickly compared to the first. In short, the process patent combined our research, other technologies, and patents to create a unique process for removing nutrients from agricultural waste streams and industrial and municipal sources.[10] By this point, we had learned what we were doing. We understood the solids part of the waste stream much better. I understood the solids part better. That really made a huge difference.

One of the most important lessons had little to do with pig poop. The first patent had a steep learning curve. The second one was more efficient, but the issue of how to expand infringement protection and when to assign rights were among our skillsets. We did not possess patent eruditions; but close. Assigning rights is a major step in the process.[11] Years ago, any research at a university funded by government monies had pronounced and appreciable obstacles.[12]

Changes in federal law have allowed researchers to personally benefit from their mental and physical efforts.[13] We were clear for many reasons. The federal government, the state of Louisiana, and the university never chipped in a dime. This was all done with private money, and we paid the university (and LSU) a fee to allow us to access their pig waste and property. The company (and subsequent companies) in the UK had all rights to the IP.[14] A quick search of Companies House[15] reveals a myriad of international businesses: Basalt Composite, Global Monitoring and Information, GVR Developments, International Waste Management Systems, KADE Associates, and Ximax Limited. The last was SSA Global.

When I asked George Miller what this meant, he informed me, "SSA Global stands for the Sword, Shield, and Armor of God. That is what you need today in this crazy world. In a world where the righteous are hit on all sides, you need SSA Global."

He may have been right.

We were going to need help from above if this project was to succeed. It did smack a little too "Crusader-like" for me. But if that was fine with the "boys" then I was OK for now.

Across the pond, though, the team was about to get a major shake-up. A dimple was about to become a pimple. Ian Selby had many ventures. He often drove fast and not just in his powder blue Bentley. He played fast and loose with the books. Selby was an accountant. He managed the books for other companies and a retirement fund. While we were busy researching, writing, and submitting a patent in California, Ian Selby was being investigated for massive fraud by the UK Inland Revenue (like the IRS in America).

He and his other business partners (no one from our group) had swindled some elderly workers out of their pensions. He had falsified expenses and levied charges that drained dry these people's life savings. Ian Selby was going to go to jail.[16] According to Court documents, the "Serious Fraud Office successfully prosecuted him for his role in the multi-million-pound fraud."[17]

I always knew he was a rascal. A likable rascal. But still a rascal.

Jim Tryand, George Miller, and I had accidently encountered Ian Selby following one of his last days in court. We were in London and saw him with his lawyer at a pub. We all stopped in to have a drink. Although the seriousness of the hearing was on his mind, he seemed impervious to what was waiting for him.

His attorney attempted to lighten the mood.

"Are you two related?" his attorney asked Jim and me.

We looked at each other. Jim was Greek. I am Scottish. I never heard anyone think that we looked alike. Our age difference should have thwarted such a question.

"Us?" I fumbled, astonished and bewildered.

George sat there laughing and enjoying another beer.

"Yeah. You could be twins. Ian, look at their ears. Their ears are the same. Fat, flat, and drooping."

I looked at Jim's ears. He looked at mine. I pinched Jim's ear like I was inspecting fruit at the market.

We started laughing.

"Maybe we are related, Jim," I suggested whimsically. I would later remark to Jim that we were 'ear twins' and we would laugh about that day in London.

But then, we all just shook our heads. Maybe we were all thinking the same thing.

Ian's imbecilic lawyer was about to get him convicted. And he did. Clearly, Selby had either committed the crime or was too close to those that had. But having a moron like this for a solicitor did him no favors. Ian was now finished and heading to jail and would not resurface for years. Before he could return to freedom, he would be ordered to pay more than $500,000 in fines to pay back in part 256 families that he robbed.[18] Regardless of my feeling for Ian Selby, I never saw him in the same light. His associates had robbed vulnerable old people. That was all that needed to be said.

Now turning to Nicholas Bristow.

Blobby.

Was he a scammer? A hustler?

Following 9/11, financial security measures increased and slowed many legitimate deals down or killed them altogether.[19] Having ties to Selby would not help. But that would come later. Blobby was the real problem. He was a nefarious scammer who claimed to be a Federal Reserve agent, Vatican money, and Morgenthau's Nazi gold in black boxes. He would certainly be a suspect on anyone's wanted list.[20] That is: he would have been. If he was a principled, trustworthy person. Nicholas Bristow was a thief, and a liar.

I was heading back to London.

But there would be another wrinkle to this project.

This one shattered the team completely.

Jim Tryand had left his family on Christmas Eve decades earlier. He literally left his car at the Detroit Airport and flew away to Geneva, Switzerland never to be heard from again by either his wife or young son. He remarried as Swiss woman (he never was officially divorced from his American wife) and lived with her for several years. Marital bliss for Jim remained a beguiling illusion. He divorced and settled into middle-age as committed bachelor.

Then, years later, his American wife died.

He had not heard from her or his son for almost 26 years. He could have left well-enough alone. Yet, in Jim's fallacious and deceitful mind, his wife's retirement or Social Security pension belonged to him. He started the paperwork necessary to claim a monthly benefit. The moment that he did this, he became visible in America. His son Peter, or Pete, showed up with a grandson. They were looking for a relationship with the man that had abandoned the family decades earlier.

Pete struck an uncanny resemblance to Jim and not just their ears. It was impossible to fail to see the genetic ties of father and son. But the years of neglect had taken their toll on Pete. He was an alcoholic and a wheeler dealer without any skills. He had some understanding of weapons and seemed to suggest that he had access to "any kind of gun you wanted." Wonderful. Now we have gunrunners in our midst. What would be next? A few years later, Pete Tryand would get arrested for illegal gun sales and wire fraud.[21]

In the meantime, Pete arranged to have dinner with me one night before I flew back to America. There was the delightful little Pub called the Yum Yum. Scrumptious grilled food, an assortment of beers, and it was cheap. We got a booth in a corner so that we could speak without being interrupted by either the wait staff or anyone else.

"John, you seem like a decent guy," Pete began.

"I hope so," I remarked. I wondered where this was going. In my experience, this is generally stated before something awful is dumped in your lap.

Then we engaged in some idle talk about skeet-shooting and hunting and camping in woods of northern Michigan. The kind of discussion Hemmingway would have retold. It was interesting but unconnected to his initial remarks.

"That is why," he continued where he had left off an hour earlier, "I want you to know what is going on."

"Ok."

This was painful for Pete. It was clear that he suffered through every syllable of the story. He told the entire story of Jim Tryand leaving the family. I sat astounded and could not speak. Then, he continued, Jim, his father, years later tried to get his dead mother's Social Security checks. His hands shook as he spoke. He was seething, but still in control. Barely.

"All I ever wanted was for my father to love me and be there for me."

I just nodded and fidgeted with my grilled peas. Then I took the top off a dark glass of ale. I needed something a bit stronger, but I also did not want to forget a word of this story.

Pete plodded on, "He would never take me to Michigan football games. My friends would go with their dads. I was stuck in some office or library until halftime, then we would go, sneak in, and find an open seat."

Pete was drunk.

Pete was pugnacious and spiteful.

Pete was about to rat his dad out.

Then it came.

He needed a little liquid fortification, and then he commenced with the narrative, "Jim's been lying to all of you. There is no Federal Reserve, or Vatican, or Portuguese oil, or Brazilian gold, or Nazi gold," Pete was one drink past sober now.

"Go on," I encouraged. Now I really needed a drink. Double Scotch? No ice?

"Those documents that 'prove' Blobby is for real," he paused, "Are all fakes. Jim is the master at document manipulation. I watched him make the documents he showed you this afternoon. He sat in his little office drawing up fake papers. My dad is a lying scammer."

I sat raddled and befuddled.

How could this be true?

We were moving forward in Belgium. Paying for permits and contracts for waste.

I pressed Pete hard now, "So where has the money come from?"

"George Miller."

"What?"

"You heard me. George Miller."

I was aghast and floored. I could not move. I remember feeling light-headed. I wanted to vomit.

George Miller had been fronting all the costs of the project from day one. Jim and Blobby had manipulated George and now he was broke. He had lost his house, his car, and his wife was divorcing him and taking the children. I knew his wife and he were having problems, but divorce? And the children? You could feel the tension in the house. By now, I had stopped staying with them when I was in London. I found a simple, humble bed and breakfast. It was clean and cheap.

Divorce? The house? The car? His savings?

It was too much to process all at once.

"George had more than a million pounds in the bank. He owned his house in Bromley. His Volvo was paid for. Are you telling me Blobby got it all?"

"Yep, and my dad helped him. I do not know how much Jim was paid, but he has orchestrated the con from day one."

I have always liked the word "obmutescent." It fits this situation well. I was dumbfounded to the point of being speechless.

"How?" I begged.

"There were processing fees, mysterious legal fees, document fees," he offered, "Then all the permits, the down payments, the leases, the trips. You name it. George paid for everything. The thing is that nothing has actually been paid for. Jim has managed all the money and kept it. SSA Global has no Belgium property, permits, waste contracts . . . nothing."

Things were making sense now.

Months earlier, an investment company, Satchel Investors,[22] had offered us almost $20 million to build the first system and to operate it in Belgium for three years. They would pay each of the directors about half a million in cash. They wanted 50% of the company. Juan and I would have received 25% of the equity and the rest evenly split between all the others. George and Jim had convinced everyone (except me) to reject the offer. The vote was 7 to 1. Now I understand why. Once the financials were disclosed, it would have been made clear that Blobby had scammed George and Jim was an accomplice. Also, if George's money was flowing, Blobby and Jim did not need outsiders.

"Jim will disappear soon. That is what he does. He has left his roommate and no longer lives in that apartment. In fact, his roommate has gone back to Prague. He is scared."

Jim's roommate was a vegan who sold prosthetics in the Czech Republic. If there is a joke somewhere in this, I have never been able to find one.

"Scared of what?"

It was obvious. Jim had scammed more than George Miller. Jim was in danger. He needed to vanish.

"He has talked about coming back with me to America. Maybe we will go to Texas. I don't know. I know this, I have a son, his grandson. Jim does not have much longer left. I would like for my son to know his grandfather. He is a shit, but he is all the family I got."

Pete was crying softly by this point. The grilled food and his sixth beer would hardly be enough medicine to cure his yearning from what would never be. He excused himself and went to the toilet. I sat there. Unable to move.

We never heard from Nicholas Bristow ever again.

Jim Tryand would disappear a few months later. Maybe he moved in with Pete for a while and they split the Social Security checks. I do not know.

In Bromley, George was booted out of his house. His wife left him and took the kids. He moved in with his brother in their mom's old house. His wife eventually divorced him. She moved back to Germany.

I was headed back to New Orleans with a heavy heart. My angst was only surpassed by my wretched disappointment.

I felt miserable.

SSA Global was in trouble, and it had barely been formed.

This story was about to really get weird.

NOTES

1. Taylor, J. B. (2007). *Global financial warriors: The untold story of international finance in the post-9/11 world.* WW Norton & Company.

2. Jervis, R. (2003). Understanding the Bush doctrine. *Political Science Quarterly, 118*(3), 365–388.

3. Ehie, I. C., & Olibe, K. (2010). The effect of R&D investment on firm value: An examination of US manufacturing and service industries. *International Journal of Production Economics, 128*(1), 127–135.

4. Brandt, J. (2007). Comparing foreign investment in China, post-WTO accession, with foreign investment in the United States, post-9/11. *Pacific Rim Law and Policy Journal, 16*, 285.

5. California Beaches. San Onofre State Beach—Nude Area. Located at https://www.californiabeaches.com/beach/san-onofre-state-beach-nude-area/. Accessed 25 June 2022.

6. Turner, K. (2000). A Review of US Patents in the Field of Organic Process Development Published during April to July 2000. And there was still lots of support for simply doing nothing. I never realized how important overcoming the desire to

just "wait and see" could be. In fact, that is always the default setting when anything new is proposed. Tilche, A., & Bortone, G. (1992). Biological purification and recovery of biogas from pig slurry. *Informatore Agrario, 48*(18 (Supplement)), 51–55. Assignment of patent rights is a tricky business. Lynch, B. C. (2012). International patent harmonization: Creating a binding prior art search within the patent cooperation treaty. *Geo. Wash. Int'l L. Rev., 44*, 403.

7. Mohaibes, M., & Heinonen-Tanski, H. (2004). Aerobic thermophilic treatment of farm slurry and food wastes. *Bioresource Technology, 95*(3), 245–254. Many were modifying existing lagoons to reduce costs and make the adoption easier. Pons, L. (2005). Blue lagoons on pig farms? *Agricultural Research, 53*(3), 14.

8. Durham, S. (2006). Floating above lagoon wastewater. *Agricultural Research, 54*(8), 11–12.

9. Apologies to the Mamas and the Papas.

10. US Patent and Trademark Office. Patent #US6692642B2. This patent also was registered through the World Intellectual Property Organization to ensure global protection. As stated in the patent, this was "A process is provided for full treatment of animal manure and other organic slurries containing suspended solids, and dissolved organic matter, nitrogen, and phosphorus, without the need for anaerobic digestion or anaerobic stabilization lagoons. The process includes centrifugal decanting, flocculation, dissolved air flotation thickening, centrate heating, acidification in fluidized bed reactor, biological removal of phosphorus, nitrogen and dissolved organic matter in microfiltration membrane bioreactor and disinfection with ultraviolet radiation. The process separates liquids from solids, converts solids into organo-mineral fertilizer pellets, and treats the liquid to dischargeable standards. Land application for effluent disposal is not required because the treatment removes all nutrients."

11. Ram, N. (2009). Assigning rights and protecting interests: Constructing ethical and efficient legal rights in human tissue research. *Harvard Journal of Law and Technology, 23*, 119.

12. Leontief, W. (1963). On assignment of patent rights on inventions made under government research Contracts. *Harv. L. Rev., 77*, 492.

13. Sample, E. A. (2018). Assigned all my rights away: The overuse of assignment provisions in contracts for patent rights. *Iowa L. Rev., 104*, 447.

14. Precedents here were developed in the software sector where employees frequently develop IP on the company's nickel. Ying, D. J. (2007). A comparative study of the treatment of employee inventions, pre-invention assignment agreements, and software rights. *U. Pa. J. Bus. & Emp. L., 10*, 763.

15. Companies House. Located at https://www.gov.uk/get-information-about-a-company. Accessed 5 February 2022.

16. Birmingham Live. Jailed for pension con. November 5, 2005. Located at https://www.birminghammail.co.uk/news/local-news/jailed-for-pension-con-14789. Accessed 25 June 2022.

17. Management Today. White-collar crime: The inside story. Located at https://www.managementtoday.co.uk/white-collar-crime-inside-story/article/550575. Accessed 25 June 2022. According to the article, which described the prison, "The institution looks out over miles of open fields, but there's nothing restful about the

place. Selby is 23 months into a five-and-a-half-year sentence for conspiring to steal £2.9 million from a company pension fund, and his daily grind as a guest of Her Majesty could not be further from the lifestyle he was once used to."

18. City Wire. Funds Insider. Pension fund fraudster ordered to pay back over £400,000. Located at https://citywire.com/funds-insider/news/pension-fund-fraudster -ordered-to-pay-back-over-400000/a323385. Accessed 25 June 2022. According to the report, "Following the confiscation order of Selby, Serious Fraud Office lawyer Gary Leong said: 'The serious Fraud Office is committed to the objective of making criminals surrender any and all benefits of crime by ensuring the confiscation of all their proceeds of criminal conduct and by ensuring that compensation is paid to their victims. We take an aggressive proactive approach, and we will use all the investigative and legal tools at our disposal to achieve that objective.'" In a similar article, Ian and his associates were called a "gang." Business Live. Gang member must pay back £400,000 over pension fund theft. December 11, 2008. Located at https://www .business-live.co.uk/economic-development/gang-member-must-pay-back-3952242. Accessed 3 July 2022. Interestingly, Ian never expressed remorse to us about this situation. He blamed everyone else in this con.

19. Levi, M., & Wall, D. S. (2004). Technologies, security, and privacy in the post-9/11 European information society. *Journal of Law and Society, 31*(2), 194–220.

20. Kaunert, C., & Giovanna, M. D. (2010). Post-9/11 EU counter-terrorist financing cooperation: Differentiating supranational policy entrepreneurship by the Commission and the Council Secretariat. *European Security, 19*(2), 275–295.

21. Times of Northwest Indiana. Peter G. Tryand, 54, of Valparaiso, is charged in Hammond federal court with buying dozens of the same expensive guns twice with checks from . . . Jul 14, 2009. Located at https://www.nwitimes.com/news/local /former-cabela-s-gun-specialist-faces-federal-fraud-charges/article_22ecb481-4b2f -5f5f-9564-604ab7cc8ed2.html. Accessed 10 January 2022.

22. This is not the name of the company. That name is being withheld because of the nature of their business. They probably would not want anyone to know that they ever considered investing in such a crazy company. This meeting also ensured that another investor would not put his money into the project. He and his wife attended the meeting, shook their heads, and went back to their "regular" company that made legitimate products. For some reason, I remember them as decent people with connections to a chocolate company.

Chapter 12

Enter Kathumi

George Miller may have lost his mind. He certainly snapped. He went so far with this project that he lost everything. His fortune, his reputation, his family, everything. Vanished. Gone. He had nothing to show for years of hard work. He was finished. He was living in his mother's old, tumble-down house with his younger brother. The placed smelt moldy and looked worse. It reeked of death and cigarettes. The wallpaper had been ripped off in most rooms exposing the rotten sheetrock; little remained of a once tidy, humble home. You felt grubby and unsanitary the moment you walked through the front door. I always felt the overwhelming desire to scratch when I entered this abode. Only a handful of old pictures and ancient-looking documents reminded you that a prominent family had once dwelled there but had fallen.

Yet more disintegration was to follow.

The team had fallen apart.

Jim Tryand and Nicholas Bristow had swindled him out of money. That was now obvious. He had lost Ian Selby due to a wild pension fraud that imprisoned him and his co-conspirators for years. Danny Streetman had walked away as well. D. Reggie and Shaky Eddie had been arrested as well. These two, in addition to selling counterfeit phones, whisky and cigarettes, they were part of the human trafficking problem of northern Europe.[1] Even before this was made official, my wife had been warned about them from a friend of ours in Brussels. Her husband worked for NATO and was watching D. Reggie. To be fair, D. Reggie had been out of the project for several months by this point. His last contribution was bringing a major Belgium environmental company to the table. The owner was eccentric. This company has since been bought out and the merger formed a significant venture. Despite several meetings, lunches, and an unusual dinner at the owner's house, nothing ever came from this.[2] They politely suggested that we look elsewhere for support. They would provide any engineering services, though, for a fee.

We were about to hit bottom.

George Miller had already hit a personal bottom.

This was a long fall from the George Miller that I knew.

A combination of financial misfortunes and personal issues saddled him with misery until he could not think straight.

His health had taken a terrible blow when shortly after Christmas in 2002 he collapsed exiting the Tube (Subway) in London. He and I were on the phone. Days earlier Joe Strummer of the band The Clash had died of a congenital heart defect. George had inherited that same condition. He simply became weak and fainted. His cell phone had gone silent. I did not find out for days what had happened. George would later claim that he had an out of body experience and flew over London when his spirit lifted out of his body. After his spirit returned to his body, George was never the same.

Many people experience a range of sensory perceptions while brushing close to death's door.[3] I have known people that claim to have seen a bright light or met their relatives "on the other side." Some recover with a profound religious commitment that enhances their faith.[4] Maybe the experience forces them to deal with issues that they have long ignored. According to some research, "People who have near-death experiences often report a subsequently increased sense of spirituality and a connection with their inner self and the world around them."[5] Perhaps the material world loses its appeal after seeing Heaven? George was now on a spiritual journey that compelled him more profoundly to "saving the planet."

His life was always devoted to some evangelical mission. Conservative Minister of Parliament Julian Lewis spoke of George, "Intellectual and visionary, liberal and anti-Communist, George Miller inspired a generation of Conservative activists in the 1980s, when the Soviet Union seemed impregnable. His operations were so extensive that few of his associates knew the full picture."[6]

George told stories. Seldom was he the hero of his stories. He shined a light on the glory of others.

Lewis would add, "[George] created the Russian Research Foundation and its Soviet Labour Review. He was a researcher and adviser to Anglo-American think-tanks, including the Institute for European Defence and Strategic Studies and the Adam Smith Institute. Later he worked with Anatoly Chubais, the Russian privatisation minister, while Yeltsin's reformers briefly held sway."[7]

Upon our fist conversations, I got the feeling that George was an intellectual. He read, wrote, and loved to research. He was fascinated by anything. He was naturally curious.

He was especially proud of his ethnic heritage. His roots were deep in the Russian soil and her politics. According to one report, George had "Descended from the Kurakin family of imperial St. Petersburg, he was born in Chile to Russian refugees who later came to England, where he was educated at Bromley grammar school (now Ravensbourne school) in south-east London. His father, Boris, worked for the Alliance of Russian Solidarists (also known as the NTS), which sought a democratic alternative to Soviet communism."[8] These credentials hung proudly on the wall of his old house.

Occasionally, when he and I had a quiet moment, we would play chess, enjoy some mint tea, and talk politics. He loved to recount the nights he spent crawling in the snows of Afghanistan carrying a Kalashnikov over his shoulder. Or smuggling Bibles and medicine to refugee camps. Doing what he could to make it better for those fighting the Soviet Union. Sometimes he organized protests to fight the Soviet system (see Figure 12.1 on the following page).

Julian Lewis further proclaimed, "For Miller the demise of Soviet Communism was an absolute certainty, provided that the West remained strong. His vision was tempered with patience and humour. He would liken the regime to an elephant, repeatedly stung between the eyes by a mosquito. The insect would be brushed aside time and again—yet, one day, without warning, the elephant would roll over, stone dead, with its feet in the air. Miller lived to see it happen; indeed, he helped to make it happen."[9] He supplied me with books and articles to read on this subject and we often had dinner with his old friends from the Thatcher days. He had an astonishing grasp of history.

When George Miller fell, the world lost a hero. That was still a few years away. He resurrected in 2003 as George Kurakin-Miller. The family ties to a dignified history was all he had. His name was venerated by those appreciative of his anti-Communist protests, riots, and governance efforts. The

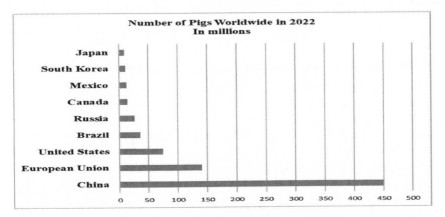

Figure 12.1. Comparison of Pig Production around the world. Source: Statistica.Com located at https://www.statista.com/statistics/263963/number-of-pigsworldwide-since -1990/. Accessed July 31, 2022. Author's calculations.

cracks were showing, though. This new George dubbed himself Kathumi.[10] A spiritual man destined to heal the world. His emails and phone calls became increasingly enigmatic and cryptic.

I had to go back to London and talk with him directly. I wanted to meet this Kathumi.

Upon my arrival, we spoke of the project at length. Then, following a late breakfast, we walked to a "healer" that George now consulted. He wanted me to de-toxify my mind, body, and spirit. Since it turned out to be a massage and a foot bath, I was OK with it. He bought me some crystals for my Chakras, and we had an organic lunch. He read some of his new age poetry as we sat near H.G. Well's Bromley (Southeast London) home, referred to as the "Atlas House." I think Billy Idol and the other "Bromley Contingents" lived close by.

As we completed our detox, poetry session, and stroll down rock-n-roll lane, he had an important favor. I assumed it would be some spiritual quest with him as Kathumi. It was nothing remotely like that. His daughter was having a party that night and he wanted me to chaperone. There would be about 40 young people, all above 18 years old, and he needed my help. I, of course, was only too glad to help.

"Am I there to watch for anyone drinking that is underage?" I checked.

"Nope. We will be providing alcohol. And food."

"So, is it cigarettes," I probed further.

"No. I don't care if they smoke tobacco or marijuana."

The new Kathumi was a bit more progressive.

"So, what is my purpose"? I wondered, needing some clarification.

"Johnnie Baby, you collect the money at the door and make sure no one gets violent. Stop anyone who tries to fight."

I was certain I could do a reasonably decent job with both. I did not need to check identification, just attitude. And collect a couple of pounds at the door.

I did.

The music was loud, and the kids were . . . kids having fun.

No violence.

We returned later that night to his brother's house. He was living there now. We hunkered down in the neglected and frazzled kitchen in threadbare chairs that I was certain would snap like tiny sticks in a dry forest at any moment. He made us some mint tea and we talked for hours. He was a bit manic, but still in control. He channeled Kathumi and spoke of his venerable principles from archaic pagan beliefs amalgamated with several mainstream religions. I felt baffled and unsettled. Kathumi was an ecumenical religious traveler.

I was demoralized and saddened. My friend was gone. The light no longer flickered is his eyes. His smile was twisted. This project had destroyed him.

"Maybe, George," I offered, "We should spend a little time discussing the pig project?"

George had found some new clients that could use his international business skills. His experience in private consulting with major international corporations paved the way for him to access emerging sectors. He still had contacts that needed his language, negotiating, and financial expertise. So, he was going to fund this project through the proceeds of another project. Rob Peter to pay Paul. It looked bad from the very start. I expressed my concerns and he continued to assure me that everything would be fine.

Despite his new age conversion, there would be no joyful investors or blissful projects.

Kathumi may have been a guru from another plane of existence, but he was not the man that George had once been. I limped back to America crushed.

Within a few months, I was back at the National Liberal Club with George Miller (a little less Kathumi but he still lurked beneath the surface). There were at least four others in the meeting; all from different nationalities. Something was not right. I felt apprehensive and nervous. The group of investors would discuss "loaning" the project money but wanted 90% equity. Miller and I (the only ones left) could split the remaining 10%. They proposed to manage the project. I was there to provide technical expertise. I was not enamored with the terms, but we were down to zero options.

They seemed to be interested in the project. Not quite like Rabo Bank but more so than the Bundesbank.

The most senior of the group raised at point, "I would like to make copies of everyone's Passport before we agree to these terms."

That is reasonable. I have found this to be ritualistic in many cultures. I was fishing through my briefcase for my passport when the next demand was made.

"I would also like a color photo of your family. A recent photo of your family."

No one moved. All just stared at me. Then George.

I was confused and on alert, "Why would you want that?"

"Because if this project fails, I want to know who to kill."

The solemn expression on his face told me he serious. A couple members of the group laughed a quick, rapid, almost like a cough laugh.

George did not. He knew this was gravely somber moment. This group did not tell amusing anecdotes or jokes. They had no use for any American stories.

"Thanks for the clarification," I thanked him, "George may I speak with you? Outside?"

George and I left the conference room and spoke quietly in the corridor. I was infuriated. I restrained myself. My neck felt hot, sweat poured from my face, and my fists were clinched. I looked like a prizefighter ready for the first bell.

Ding!

"George, are you crazy? What have you gotten yourself into? What have you gotten me into? Pictures of my family?"

"Johnnie Baby, it will be fine. This will fund the entire project. Please, do not embarrass me. I have seen through the eyes of Kathumi and all will be well."

I explained how he had lost his mind and that I was catching a taxi, getting my belonging, and heading back to America. If he wanted to take his life into his own hands, that was his business. I had a special responsibility to my family to not put their lives in danger to fund this hog wash. My commitment to protect my family meant something to me. He may have had supernatural powers as Kathumi but I was certain that I was a pure mortal.[11]

George had learned nothing from losing his family, house, and all his money.

He was wretchedly desperate. At one point, George spent some time in Oregon with other similar devotees of Kathumi. They wanted to establish a "water-based society" for the purification of the soul. This included buildings in the shape of the Star of David and a vegan diet. I never understood what any of that meant. I was just wanting to clean up pig waste. Still, he was not fearful nor disheartened.

He was foolhardy and reckless. This is how people lose more than their money; they lose their lives.

He was drowning and could take me under with him.

There had to be other options.

Nothing was left but hog wash. This time it was not a fat man from Canada. It was through a mystical holy man.

Kathumi possessed George Miller.

NOTES

1. Derluyn, I., & Broekaert, E. (2005). On the way to a better future: Belgium as transit country for trafficking and smuggling of unaccompanied minors1. *International Migration, 43*(4), 31–56.

2. The owner of the company had a life-sized Rhinoceros in his office. It was made from copper. It was very imposing as it was looking ready to charge when you stepped out of the elevator. He was a homeopathic guru. Once during lunch, he pulled a small strand of hair from the back of my head. Then, he explained that I had a magnesium deficiency and gave me a loaf of whole grain bread, some dark elixir to drinks, and a packet of vitamins. That turned out to be the least insane moment. He also was

devoted to never smelling anything foul. Through out his gardens there were nozzles emitting fragrances upon demand. As I am fond of gardenia, I did not mind this.

3. Roberts, G., & Owen, J. (1988). The near-death experience. *The British Journal of Psychiatry, 153*(5), 607–617. More interesting books are as follows. Ring, K., & Valarino, E. E. (2006). *Lessons from the light: What we can learn from the near-death experience*. Red Wheel/Weiser.

4. Fox, M. (2003). *Religion, spirituality, and the near-death experience*. Routledge. Greyson, B., Holden, J. M., & James, D. (Eds.). (2009). *The handbook of near-death experiences: Thirty years of investigation*. Praeger.

5. Khanna, S., & Greyson, B. (2014). Near-death experiences and spiritual well-being. *Journal of Religion and Health, 53*(6), 1605–1615.

6. *The Independent*. George Miller-Kurakin: Anti-communist campaigner who inspired Conservative activists during the Cold War. November 26, 2009. Located at https://www.independent.co.uk/news/obituaries/george-millerkurakin-anticommunist -campaigner-who-inspired-conservative-activists-during-the-cold-war-1827401 .html. Accessed 26 June 2022.

7. Ibid.

8. *The Guardian*. George Miller-Kurakin obituary. February 12, 2010. Located at https://www.theguardian.com/theguardian/2010/feb/12/george-miller-kurakin -obituary. Accessed 26 June 2022.

9. *The Independent*. One of my favorite stories about George involved him risking his life to save some troops. "Exploiting a distant Muslim family link, he entered Afghanistan to negotiate the release of captive Russian soldiers who had survived by converting to Islam." That was the way he was. Fearless, and brave. One of the books that George gave me has been a resource of mine ever since. Griffin, G. E., & Bramhall, M. (1998). *The creature from Jekyll Island: A second look at the Federal Reserve* (p. 608). American Media. I have read and re-read this book many times. A little conspiracy for me. Not as bad as others. Greider, W. (1989). *Secrets of the temple: How the Federal Reserve runs the country*. Simon and Schuster.

10. Kathumi. Located at http://vixen-awakent-primary.hgsitebuilder.com/kathumi. Accessed 3 July 2022. According to the Kathumi website, "Formerly Chohan of the Second Ray of Divine Illumination, Master Kuthumi (also called Koot Hoomi and K.H.) now serves with Master Jesus/Sananda as World Teacher. One of his earlier incarnations was as Pythagoras, (circa 582–circa 507 BC) the Greek philosopher and mathematician. He also incarnated in the first century as Balthazar, one of the three Magi (Wise Men) who journeyed from the East to pay homage to the Christ Child." George Miller had snapped. The pressure from home and the project had worn him down.

11. George began channeling Kathumi during one of my visits. I was raised in a charismatic Christian home and the entire concept of channeling was akin to evil spirits possessing someone. I was unaware that there was an entire culture of Kathumi channelers out there. Here is one such person. YouTube. Located at https:// www.youtube.com/watch?v=8fPFd0Wj56A&list=PL23BD27207D9CEA1B&index =1. Accessed 3 July 2022. Mostly these sessions are filled with a hodge-podge of mysticism and vague language about past lives and generic love-peace-environmental

stuff. I find this to be very sad and tragic. George Miller travelled to a retreat with these people in Oregon to undergo some type of "healing." It involved water and air. I am in favor of both. This is just New Age nonsense about time-shifting and bringing about making the world better. Harmless until you believe. Too many people fall for such twaddle in desperation. There is an older book where much of the content has been recycled with a little more contemporary information: Sinnett, A. P. (1924). *The Mahatma Letters to AP Sinnett from the Mahatmas M. & KH* (Vol. 2). TF Unwin Limited. This seems to be the origins of the Kathumi craze. And a more recent book: Mills, J. (2014). *Reflections on an ageless wisdom: A commentary on the Mahatma Letters to AP Sinnett*. Quest Books.

Chapter 13

The Chlorine Mafia

This project was all but officially dead. The next couple of years were difficult. None of the original members of the team were left. Even Juan was gone by this time. George Miller spiraled down further, and he eventually had to accept governmental charity to buy food. He had lost a lot of weight and his face was a haggard shell of his radiant past.

But he was still fighting for the project. He had taken the work with the group we had met at the National Liberal Club. I assumed it was going badly and did not ask any questions. I am sure he felt my indignation and exasperation with his international business partners. George had made a small splash internationally as he assembled a team to take-over Gazprom.[1] This is the national oil company of Russia worth billions of dollars. George could not pay rent. His delusions exceeded Icarus' dreams of flight.

He reached out to (probably) the only one that he could: Ian Selby. He had been recently paroled and needed some legitimate work to justify his early release by Her Majesty's Government. Following high profile fraud cases like Selby's, the UK had passed the *Pensions Act* (2004) to afford greater powers of investigation for the police and improved financial recompensating for victims.[2] Not everyone was happy about the early release of the fraudsters that had ripped off old age pensioners.[3] None of the victims received full restitution. Now Selby was free.

The meeting was held in a warehouse in Ealing (West London; film district). A Tunisian restaurant catered our meeting. The food was incredible. There, a group of local businessmen accompanied by Selby and his cellmate had invited a Dutch engineer (whose name was impossible for me to pronounce) to discuss the prospects for a chemical disinfectant that could be part of the SSA Global repertoire of products. No one had any environmental science, policy, technology, or regulatory experience.

Except for me.

Normally, I could have relied upon Juan to guide me through the murky waters overflowing with pig slurry. No more safety net. Now, I was on my own. I was there to offer an opinion about the viability of the product. Based upon my recommendation, there would be an investment into a new business.

There was a long lecture about the product. The parent company developed a powdered version of a gas that had been used as a bleach substitute. It was a chlorine dioxide (ClO2) compound that acted as an oxidizer to destroy or remove (through filters) bacteria, viruses, and algae from water.[4] It could supplement the patented technology (both the device and the process) and offer a wider range of treatment options. The discovery of CLO2 had occurred in 1814 by Sir Humphrey Davy when he combined sulphuric acid (H2SO4) and potassium chlorate (KClO3). Later, he would substitute the sulphuric acid with hypochlorous acid (NaClO3). Davy made it possible to produce a stable, high-concentration product that could dissolve in water.[5]

So why had this miracle compound never replaced chlorine? Was there some improper or nefarious force at work here? I knew next to nothing about CLO2.

He explained that it took many years of research and product development to reduce the dangerous side-effects, especially for public health,[6] and the risk of explosions associated with the product.[7] When mixed at high concentrations the compound will simply detonate and kill people. CLO2 had decades of research backed up claims of the compound being superior to chlorine.[8] More recent studies had discovered additional disinfection applications for cooling towers and drinking water.[9] And, unlike chlorine, CLO2 has been proven to have almost no trihalomethanes (THMs) in drinking water.[10] Even more documentation established CLO2 as able to precipitate iron[11] and magnesium[12] from water. Adding a sand filter brightened the water for reuse as potable.

It seemed that I was getting further and further away from hog slurry.

Pig waste was fine.

All agreed.

That could be a part of the business.

These people wanted to do more.

"So, what is the problem?" I was becoming impatient.

"The problem," he explained between an endless chain of cigarettes, and the nervous tick of looking around, "Is the Chlorine Mafia."[13]

"The what?" Like a choir, we all asked at the same time, interrupting his train of thought.

He repeated it.

Like it was a shibboleth for an incantation for a secret organization or secret knowledge.

Then, he spoke is more hushed tones, still looking around and smoking, "When I was in South Africa, they tried to kill me. My houseboat in Amsterdam was looted and set on fire. These people are serious. My girlfriend has gone into hiding in fear for her life. These people make billions and do not want us taking away their money."

Whether there was a Chlorine Mafia or not, this (not quite flying) Dutchman believed it. Maybe some of the people at the table believed it.

"I am intrigued," I offered, "I know little about the product, the Chlorine Dioxide compound, and I would like to tour your facility, take some samples, and conduct my own laboratory tests."

He was adamant, "No way."

Finally, he yielded grudgingly. I could have a few samples with some basic instructions but no tours. None of us could visit the plant. They could not trust anyone with their location. We might try to steal the patent or sabotage the facility. I was thinking about how much I missed Blobby while he spoke.

George Miller was not impressed.

He still hoped that his new business friends would allow him to invest in a full-scale plant in Belgium. That dream had vanished the moment they asked me for pictures of my family. I would never accept money on those terms. George was simply in too deep to turn back now. He remained assiduously zealous.

Selby and his new business partners wanted to meet with me later that month. I would have time to conduct research and test the product. Mainly, I investigated the concentration levels. If the product could produce a concentration level like they claimed, maybe we had another line of work. His offices were in Brentwood (Essex County; East of London). Modest, functional, and unpretentious. Working class offices. Not the City of London. Far away from the National Liberal Club. George and I came separately, and he showed surprise that I was back in the UK without his knowledge. I could not stay in his brother's house again. I found a small Bed and Breakfast in Brentwood that was close to a Fish-N-Chip shop and an Off-Track-Betting Parlor. I needed to put some distance between us. As George Miller, he was an asset; as Kathumi, he was a deranged loon. When he arrived, I was mournful and morose to see what had become of my old friend. George was rail thin, and his ashen face proved his exhaustion. His beard was poorly trimmed, a little greyer, and his shirt was covered in food and grease stains. The office staff laughed at him. One office worker suggested that George was in a "food army" since he has so many medallions from past gastronomic campaigns. I was hurt.

George had a beautiful spirit. To see him treated so uncharitably was agonizing.

Then he made it worse.

George suggested that we spend $100,000 on an economic analysis study to determine the market demand for chlorine and CLO2. Then, hire some universities to conduct a series of tests across several waste streams.[14] Normally, with an unlimited supply of money, this may have been the best approach. But considering the product was a proven oxidizer in the water disinfection sector, there was already enough data to decide.[15] There was also evidence supporting complementary uses with other products, such as ozone, for oxidation and disinfection.[16] This mean we should consider CLO2 as another technology to treat wastewaters across other sectors.

Our Dutch partner was right about one thing: there was a global chlorine regime, but not a mob.[17] Chlorine was a mature industry with 1,,000s of companies, uses, and decades of data on applications as diverse as water treatment to the bleaching of paper and pulp.[18] World demand eclipsed 50 million tons annually.[19] Yet, like any chemical, there were risks associated with over usage. The World Health Organization (WHO) and the Food and Agriculture Organization (FAO) raised awareness in their latest report about too much chlorine residuals, especially mercury, in food and water.[20] Others found toxic levels of chlorine in consumer products.[21]

In terms of finding an application for pig waste, many wastewater engineers, whose background was municipal sewage, recommended chlorine disinfection for lagoons. Research supported that in the 1960s.[22] Once again, Chinese scientists were ahead of everyone here. They noticed bioaccumulation of chlorine in the blood of pigs, and too much in the water; thus, impacting the surface waters and soil.[23]

But if we had a chlorine alternative?

One that was cost-effective?

And safe for workers, the environment, and the pigs?

Easy to administer?

Maybe our project was not dead.

During World War Two, researchers had found even the nastiest water could be purified with filtration and CLO2.[24] Maybe we could really close the ecological loop and have zero discharges: the goal of every water treatment system.[25] There were certain challenges, especially with struvite (that salty stuff excreted by sweaty hogs), and some minerals, but if we could recirculate the liquid waste stream, recycle and reuse (for fuel) all the solids, and neutralize all odors, then the pig project was bolstered to viability.[26]

Could George Miller's pal, and ex-con Ian Selby rejuvenate our pig project? I left the UK more optimistic than I had been in years.

Or maybe I was just being foolish and gullible (again).

This time I demanded proof. I wanted to see firsthand how this new product worked. I was recalled to the UK to conduct an experiment disinfecting the liquid portion of compost, or leachate. Analysis showed high nutrient content and some residual metals and bacteria.[27] If this leachate could be disinfected and rendered safe, there were uses for it. It could be combined with sewage sludge and made into fertilizer.[28] The pig waste research had provided tons of data on comingling waste streams. Maybe we could mix compost, sewage sludge, and CAFO waste, treat the water through our patented system, disinfect the leachate using the CLO2, and produce a fuel pellet from the biogas. That would work.

There could there be life left in this project.

Along with a few of the new Selby gang, I trekked to a compost facility in south England. It received the typical woody, green, and sewage sludge wastes. Also, it applied chicken waste (high nutrient, and ammonia) once the material was blended. The goal was to produce a soil amendment for landscaping.[29] The composting system was an upright vessel, like a tube, where the waste stream was fed into the top and the leachate filtered through the bottom. It would then be collected, treated, and disposed or reused.

My "researchers" did not like this application from the beginning. It was too nasty. It smelt. They got dirty. Once I discovered that none of them had any technical competence, experience, or curiosity regarding CLO2, I felt that weird sense of doom. The most "technical" person turned out to be a former hair stylist. She understood chemicals. The rest were salesmen. And not good ones. I was supposed to be the technical expert.

There was no need for a Chlorine Mafia. This group would sabotage itself.

The experiment was a total failure. There was plenty of research on the use of CLO2 for odor reduction at compost facilities.[30] Most studies suggested that pre-treatment of the water before it was in contact with the organic waste was key.[31] This system would have to be totally redesigned ands the waste streams separated. That meant costs. Those costs were greater than doing nothing. They did nothing.

Maybe the failure could provide some insight on possible uses?

Now I had experience with this new product. I could apply this to the pig waste sector. CLO2 had been used for decades to reduce odors in drinking water.[32] More recent studies proved that CLO2 had less side-effects than traditionally used disinfectants.[33] I would have to dive deep into the literature. Also, I needed another experiment to build up my résumé. I did not have to wait long.

If this pig project was ever going to get funded, then it was time for me to push forward. I briefed George Miller about the possible uses of CLO2 for our project. Again, he was not supportive. He believed his "friends" would finance the project and wanted me to hold out until the money came through. I could not.

Then, fate struck again.

The 2007 Financial Crisis hit the world markets like Hurricane Floyd had smacked those pigs around in North Carolina.[34] Whatever scheme George had been cooking up had boiled over and spilt on the floor of his mother's dirty old house. He was ruined. His clients would start asking questions about their "investments." His staggering rates of return could no longer be guaranteed.[35] These people were not interested in hearing about Icelandic Banks and London Public Works accounts. They wanted their money. Now.

Figure 13.1.   Banana Plantation Research, where author begins developing other projects. Courtesy of the Author.

George Miller was sinking and even Master Kathumi and all the other mystical chants and readings could not save him. I grieved for my friend. He was spiraling even further down, and I could offer him nothing. Ian Selby wanted nothing further to do with him. That meant whatever I discovered for CLO2 would not include George.

What could I do?

I went to Costa Rica to test the new product on fungus and rot on bananas (see Figure 13.1). I found that CLO2 worked extremely well in combatting fungal growth on bananas and other tree fruits.[36] Again, Chinese researchers had been ahead of the rest of the world.[37] When a full disinfection of the facility occurred (cutting, hauling, washing, wrapping, shipping), the reduction contamination was extraordinary. It cut losses by more than half. The uses of CLO2 were beginning to pile up. Costa Rica was a major success. I would retune several more times over the next few years and expand the use of CLO2 from bananas to mangoes to pineapples. I was even able to take my family to vacation (while I worked). It certainly was a fun time.

But was I getting any closer to an application for pig poop?

More hog wash?

I needed to get back on track or walk away.

How could I just walk away from years of research? Personal sacrifice? The hopes of so many? How could I leave George Miller behind? I needed time to think through all my options. I needed to decide. Sometimes the most prudent thing to do is admit defeat and move on.

Yet I held out hope.

NOTES

1. Gazprom. Located at https://www.forbes.com/companies/gazprom/?sh=253380e57ec8. Accessed 30 June 2022.

2. Pension Act of 2004. Located at https://www.legislation.gov.uk/ukpga/2004/35/contents. Accessed 26 June 2022.

3. AccountingWeb. Regulator criticised in £3 million pension scheme theft. November 10, 2005. Located at https://www.accountingweb.co.uk/business/finance-strategy/regulator-criticised-in-ps3-million-pension-scheme-theft. Accessed 26 June 2022.

4. Deshwal, B. R., & Lee, H. K. (2005). Manufacture of chlorine dioxide from sodium chlorite: Process chemistry. *Journal of Industrial and Engineering Chemistry, 11*(1), 125–136.

5. Lentech. Located at https://www.lenntech.com/processes/disinfection/chemical/disinfectants-chlorine-dioxide.htm#:~:text=Chlorine%20dioxide%20was%20discovered%20in,large%20quantities%20of%20chlorine%20dioxide. Accessed 8 March 2022.

6. Lardieri, A., Cheng, C., Jones, S. C., & McCulley, L. (2021). Harmful effects of chlorine dioxide exposure. *Clinical toxicology* (Philadelphia, Pa.), *59*(5), 448.

7. Jin, R. Y., Hu, S. Q., Zhang, Y. H., & Bo, T. (2008). Research on the explosion characteristics of chlorine dioxide gas. *Chinese Chemical Letters, 19*(11), 1375–1378.

8. Benarde, M. A., Israel, B. M., Olivieri, V. P., & Granstrom, M. L. (1965). Efficiency of chlorine dioxide as a bactericide. *Applied Microbiology, 13*(5), 776–780.

9. Gordon, G., & Rosenblatt, A. A. (2005). Chlorine dioxide: the current state of the art. *Ozone: Science & Engineering, 27*(3), 203–207.

10. Lykins Jr, B. W., & Griese, M. H. (1986). Using chlorine dioxide for trihalomethane control. *Journal-American Water Works Association, 78*(6), 88–93.

11. Knocke, W. R., Van Benschoten, J. E., Kearney, M. J., Soborski, A. W., & Reckhow, D. A. (1991). Kinetics of manganese and iron oxidation by potassium permanganate and chlorine dioxide. *Journal-American Water Works Association, 83*(6), 80–87.

12. He, Z., Jahan, M. S., & Ni, Y. (2009, October). Using Magnesium Hydroxide for pH Control to Improve the Chlorine Dioxide Brightening Stages in ECF Bleaching Sequences. In TAPPI Engineering, Pulping & Environmental Conference.

13. Kremyanskaya, E. A. (2011). Cartels in Russia: Fight chronicles: Way to success. *New Journal of European Criminal Law, 2*(4), 426–439.

14. Gordon, G. (2001). Is all chlorine dioxide created equal? *Journal-American Water Works Association, 93*(4), 163–174. The answer to this question would take me several years to figure out. Of course, the answer is no.

15. Knapp, J. E., & Battisti, D. L. (2001). Chlorine dioxide. *Disinfection, Sterilization, and Preservation, 5*, 215–228.

16. Long, B. W., Hulsey, R. A., & Hoehn, R. C. (1999). Complementary uses of chlorine dioxide and ozone for drinking water treatment. *Ozone: Science & Engineering, 21*(5), 465–476.

17. Gilliatt, B. (2003). Operating in contested environments: The experience of the chlorine industry. In *The Challenge of Change in EU Business Associations* (pp. 123–135). Palgrave Macmillan, London.

18. Stringer, R., & Johnston, P. (2001). *Chlorine and the environment: An overview of the chlorine industry.*

19. Tundo, P., He, L. N., Lokteva, E., & Mota, C. (Eds.). (2016). *Chemistry beyond chlorine.* Switzerland: Springer.

20. World Health Organization. (2009). Benefits and risks of the use of chlorine-containing disinfectants in food production and food processing: report of a joint FAO/WHO expert meeting, Ann Arbor, MI, USA, 27–30 May 2008.

21. Winder, C. (2001). The toxicology of chlorine. *Environmental Research, 85*(2), 105–114.

22. Irgens, R. L., & Day, D. L. (1966). Laboratory studies of aerobic stabilization of swine waste. *Journal of Agricultural Engineering Research, 11*(1), 1–10.

23. Zhang, C., Du, Y., Tao, X. Q., Zhang, K., Shen, D. S., & Long, Y. Y. (2013). Dechlorination of polychlorinated biphenyl-contaminated soil via anaerobic composting with pig manure. *Journal of Hazardous Materials, 261*, 826–832.

24. Synan, J. F., MacMahon, J. D., & Vincent, G. P. (1944). Chlorine dioxide: A development in treatment of potable water. *Water Works & Sewerage, 91*(12), 423–6.

25. Goldblatt, M. E., Eble, K. S., & Feathers, J. E. (1993). Zero discharge: What, why, and how. *Chemical Engineering Progress*;(United States), *89*(4).

26. Ong, H. K., Choo, P. Y., & Soo, S. P. (1993). Application of bacterial product for zero-liquid-discharge pig waste management under tropical conditions. *Water Science and Technology, 27*(1), 133–140.

27. Romero, C., Ramos, P., Costa, C., & Márquez, M. C. (2013). Raw and digested municipal waste compost leachate as potential fertilizer: Comparison with a commercial fertilizer. *Journal of Cleaner Production, 59*, 73–78.

28. Chatterjee, N., Flury, M., Hinman, C., & Cogger, C. G. (2013). Chemical and physical characteristics of compost leachates. A Review Report prepared for the Washington State Department of Transportation. Washington State University.

29. Hue, N. V., & Sobieszczyk, B. A. (1999). Nutritional values of some biowastes as soil amendments. *Compost Science & Utilization, 7*(1), 34–41.

30. Hentz Jr, L. H., Murray, C. M., Thompson, J. L., Gasner, L. L., & Dunson Jr, J. B. (1992). Odor control research at the Montgomery County regional composting facility. *Water Environment Research, 64*(1), 13–18.

31. Mussari, F. P., Smith, J. E., & Jr, D. M. (2013). Accelerated composting methods and equipment. *European Journal of Marketing*.

32. Mounsey, R. J., & Hagar, M. C. (1946). Taste and odor control with chlorine dioxide. *Journal-American Water Works Association, 38*(9), 1051–1056.

33. Gordon, G., & Rosenblatt, A. A. (2005). Chlorine dioxide: The current state of the art. *Ozone: Science & Engineering, 27*(3), 203–207.

34. Acharya, V., Philippon, T., Richardson, M., & Roubini, N. (2009). The financial crisis of 2007–2009: Causes and remedies. *Restoring Financial Stability: How to Repair a Failed System*, 1–56.

35. Flannery, M. J., Kwan, S. H., & Nimalendran, M. (2013). The 2007–2009 financial crisis and bank opaqueness. *Journal of Financial Intermediation, 22*(1), 55–84.

36. Roberts, R. G., & Reymond, S. T. (1994). Chlorine dioxide for reduction of postharvest pathogen inoculum during handling of tree fruits. *Applied and Environmental Microbiology, 60*(8), 2864–2868.

37. Wen, G., Xu, X., Huang, T., Zhu, H., & Ma, J. (2017). Inactivation of three genera of dominant fungal spores in groundwater using chlorine dioxide: Effectiveness, influencing factors, and mechanisms. *Water Research, 125*, 132–140.

Chapter 14

Things Fall Apart

According to Victorian-era British Prime Minister Benjamin Disraeli, "Change is inevitable."[1] How true that is. Especially for academics. It is not uncommon for professors to relocate early in their careers. You go where there are jobs. Not everyplace has a university. I have heard professors described as "nomads" with books. So, we were about to move again. Not to Bruges, Bromley, or Brentwood. I was fortunate. Because I had taught, published, and researched, I had more flexibility in getting a job at another university.[2] Many newly minted PhDs have yet to accumulate such a CV.

Now that our second child was born, we decided to move to my wife's hometown of Lafayette. To be closer to her family. Just a couple of hours west of New Orleans, this was not a terrible burden. I managed to still teach at Tulane University (three days per week) and consult on the pig project. Since I had completed my last environmental documentary film series, I was no longer employed at the University of New Orleans' Urban Waste Management & Research Center (UWMRC).

More change was coming.

I was informed that my working-class, non-designer PhD would never allow me to be anything other than a popular professor, teaching popular courses, but certainly would never be granted tenure or promotion beyond assistant professor at Tulane University. So, I decided to look elsewhere in Louisiana for a professorial gig. I quickly found a more welcoming job in the northern part of the state at the University of Louisiana Monroe (ULM). It is a decision that I have never regretted. And, within a year of moving to Lafayette, I found myself commuting and eventually living in Monroe while my family continued behind. Since I had spent so many years on the road, especially pursuing pig poop projects and directing environmental films, it really was not a heavy load.

An interesting note to the transition from New Orleans to Monroe. I left Tulane University in May 2005 and started teaching in Monroe in August. That August, Hurricane Katrina swallowed the Big Easy and breached levies causing untold suffering across the Gulf Coast. The professor I replaced had moved to southern Mississippi and his university was wiped out.

Karma? Luck? Born under a Good Sign? All of that!

As I settled into my new surroundings, I was soon approached by some Minnesota businessmen having problems with animal waste and that door seemed to show the way forward. The company had access to lots of cow and hog manure and their own proprietary technology (compost filters for agricultural waste). I travelled to Minnesota a few times, enjoyed the lake, ate some Walleye (fish), and made some lasting friends.

Then, I was contacted by an Iowa company that needed some regulatory and compliance help to ensure their permits would be granted.[3] Also, they had some technical issues with their process that seemed to be sufficiently complex requiring my 'expertise.' I spent more than two years going back and forth to Des Moines and other parts of the state. I did a ton more consulting and made even more friends.[4]

My main focus remained pigs. I knew this. I needed to refocus.

That is when I went back to the UK to meet with Ian Selby. It seems all twisted and contorted because it was. George Miller was grasping for anything to keep the project afloat. Still, we had little to show for our efforts but more hog wash. It was collapsing and there was nothing I could do to save it.

But I have excelled at making lemonade from lemons throughout my career. This would be my greatest "silver lining" quest. I needed to turn pig crap into gold. As such, I was gaining experience and a reputation for being an environmental expert. And not just with pig slurry. Across all sectors. I accelerated obtaining certifications across many sectors: wastewater, drinking water, landfilling, recycling, composting, Brownfields, solar panels; anything environmental to improve my technical understanding to the subject.

Additionally (see Chapter 13), I was picking up some knowledge and applications for the CLO2 and had conducted experiments in the field. I even had a pilot project followed by a full demonstration. I had gone to Costa Rica and found a useful purpose that was cost-effective. I was, though, getting further and further from my initial goal: to solve the pig waste problem.

The year 2009 was a crossroads for many Americans and others around the world. Especially coming out of the Financial Crisis of 2007. I never felt at ease or relaxed for too long. There seemed to be good followed by bad followed by good . . . Barack Obama was inaugurated as President. Then, Michael Jackson died mysteriously. There was a ceasefire in the Gaza Strip. And Captain "Sully" landed US Airways Flight 1549 in the Hudson. Awareness regarding Climate Change culminated in the First International Day of Climate Action. These events captured the headlines.[5] Emotionally, I was dragged apart.

Personally, the year 2009 was filled with devastation followed by catastrophe.

On January 3rd, my last remaining brother (Paul) died at the age of 46.[6] My oldest (Lewis) and youngest (James) brothers had died years before. Paul and I had not spoken for several months for some ridiculous reason that seemed important at the time. Unfortunately, neither my father nor I were not allowed to participate in the memorial service. As he was my favorite brother, this has always grieved me.

Then, on April 25th, I was informed that my dad (Philip; "Pastor Buzzy") had passed away.[7] My dad and I became closer after my mother (Tommie; Sister "Ta Ta" as her friends called her) died about a decade earlier. I was asked to officiate the services with a couple of my dad's preacher friends. I must have done a good job as I have "preached" numerous funerals ever since. Since my dad died intestate, I spent the next several months working out the terms to settle his estate. I tried to balance honoring what I thought were his final wishes and doing what was fair for those who remained.

There was less and less contact between George and I over the next six months. The project seemed to be languishing. It had deteriorated to the point of desperation. George was still hopeful, of course. That was his nature. The eternal optimist. But he was withering away. Our conversations finally stopped. There was little to say.

Then, George Miller-Kurakin died October 23, 2009. I was overwhelmed with grief. Maybe it was the accumulation of deaths, but I could barely move for a week. The hardship was made worse because of our distance both geographically and emotionally. I found my thoughts drifting to happier times in France, Belgium, the Netherlands, and the UK. We enjoyed so much vodka, wine, tea, bread and cheese . . . and stories . . . the stories are what I missed the most.

One of my favorite moments was when George and I traveled to Paris to meet with one of the most successful, well-known law firms in the world. The meetings were to secure our patents and licensing agreements outside of America. After our meetings, we leisured along the Seine River like tourists. We ate baguettes, cheese, and drank wine. Too much wine. We talked with locals and sang a Pink Floyd song with an Albanian street performer. I do not think I have ever heard that song since without thinking about that day in Paris.

It was a special moment.

George had lived a life that meant something to the freedom of millions of people that would never know his name. The following year, his brother sorted through George's work and made a substantial donation. Today, his writings, research, and collections are housed at the Hoover Institute at Stanford University.[8] I always like this description, "Except for the fact that his suits came from Oxfam [a UK charity], George—bearded and with the social ease of a Russian aristocrat—could have stepped out of a novel by Tolstoy."[9] Recent analysis points to a time when George Miller risked it all for his fellow Russians. Using corporate and university contacts, he worked tirelessly to overturn the oppressive Soviet system.[10]

Once the collapse of the Soviet Union was eminent, George and his father Boris travelled back to Russia to help form the new (hopefully democratic) government. Unfortunately, he and Boris were both passionately on opposite sides. According to reports, "George argued powerfully that the historic opportunity should not be missed to help shape Russia's democratic future. An opposing faction, which included his father, argued that the offer was a trap set by its old KGB enemies and that it would be better to wait for a more propitious moment to enter government; perhaps both proposals contained an element of wish-fulfilment. Boris's faction won by a single vote, resulting in a lasting rift between father and son, and George's immediate resignation from the NTS [the counter-revolutionary National Alliance of Russian Solidarists]."[11]

George seldom spoke about his father. I understood all too well. Sometimes those that you love the most are the ones that cause you the most pain.

There have been countless attempts to explain George's untimely demise. He was only 54 years old. Was there foul play? Could his business partners have rubbed him out? Old enemies from his anti-Soviet days? I have spoken many times to investigative agencies in the UK regarding his death. The official cause of death is listed as heart failure. That is probably right. It is as good of an explanation as any.

A broken heart doomed him after a decade of failure, loss, and sacrifice took its toll.

Even his strange attraction to the teachings of Kathumi seemed to make better sense in hindsight. He wanted to change the world . . . again. He wanted to live for something greater than a paycheck.

His death was the final nail in a coffin lined with the dreams of dozens of people intended for the benefit of millions of people. His death signaled the end to back-breaking research, millions of dollars in research, and countless meetings and presentations. George had held this unkempt and unrepentant cluster of zealots together. To be fair, before his body was laid to rest, this project was all but over. In hindsight, things always seem to be easier to untangle.

SSA Global would file for bankruptcy and be dissolved on June 29, 2010.[12]

The pig project was now dead.

Officially dead.

And so was George.

NOTES

Chapter title is not in reference to the late Chinua Achebe's debut novel of that name (1958). All apologies to him. It is still one of the most compelling tragedies ever.

1. Benjamin Disraeli. October 29, 1867. Located at https://www.oxfordreference .com/view/10.1093/acref/9780191826719.001.0001/q-oro-ed4-00002812. Accessed 12 May 2022.

2. Bozeman, B., & Gaughan, M. (2011). Job satisfaction among university faculty: Individual, work, and institutional determinants. *The Journal of Higher Education, 82*(2), 154–186.

3. Codner, S. (2000). Varied organic feedstocks sell opportunity in Iowa. *BioCycle, 41*(8), 30–30. The company was featured in this article. They had been part of an Iowa grant and research project that had turned waste into compost.

4. The diversity of food in Iowa is amazing. Çela, A., Knowles-Lankford, J., & Lankford, S. (2007). Local food festivals in Northeast Iowa communities: A visitor and economic impact study. *Managing Leisure, 12*(2–3), 171–186. Besides the porkchops, there are many ethnic restaurants from assorted countries. Great Dutch bakeries.

5. The People History Page. What happened in 2009. Located at https://www .thepeoplehistory.com/2009.html. Accessed 30 June 2022.

6. Find a Grave. Located at https://www.findagrave.com/memorial/32680620/ paul-d-sutherlin. Accessed 30 June 2022.

7. Find a Grave. Located at https://www.findagrave.com/memorial/36481328/ phillip-e_buzzy_-sutherlin. Accessed 30 June 2022.

8. Online Archive of California. Overview of the George Miller-Kurakin papers, 1949–1990. Located at https://oac.cdlib.org/findaid/ark:/13030/kt6k4038p5/entire _text/. Accessed 12 March 2022. According to the abstract, this includes: "Correspondence, writings, petitions, leaflets, flyers, clippings, other printed matter, and photographs, relating to clandestine activities of the Narodno-Trudovoi Soiuz in the Soviet Union and its public activities in the West, in opposition to Soviet communism and in support of Soviet dissidents. Includes papers of Boris Miller, Narodno-Trudovoi Soiuz leader and father of George Miller-Kurakin."

9. Frost, Gerald. George Miller: Anti-Communist. *The Critic*. August 22, 2021. Located at https://thecritic.co.uk/moscow-coup-george-miller-the-anti-communist/. Accessed 30 June 2022. The author adds, "He was born in Chile in 1955. His father Boris, an engineer, had migrated from Serbia where his own father, a White Russian émigré, had been murdered in front of the family. So, began a pattern of events in which politics shaped the lives and hopes of three generations of the Miller family." George Miller recruited in 60 countries to establish an anti-Soviet infrastructure of ideas and people.

10. Buchanan, Kirsty. The senior Tories with the secret Soviet past. *The Telegraph*. March 19, 2021. Located at https://www.telegraph.co.uk/news/2021/03/19/senior -tories-secret-soviet-past/. Accessed 30 June 2022.

11. Frost. The author would add, "Boris died penniless in a Moscow hospital following a heart attack in 1997. He had spent his last years as the Russian head of an international human rights body, appearing regularly on Russian television to denounce various human rights violations." I knew that this always bothered George.

12. Companies House. SSA Global Prospects. Located at https://find-and-update .company-information.service.gov.uk/company/02407584. Accessed 30 June 2022. George had created several companies with the SSA Global name. All dissolved immediately upon his death or were wrapped up with a year. Jim Tryand had long since vanished and the UK government was forced to cleanup the mess that was left behind. None of these companies had any assets or bank accounts. All companies were flat broke by this point.

Conclusions

Any given project will probably fail. Failure was always the most likely outcome for this project. A pig poop invention embraced by the same people content on despoiling the environment would never work. Yet a courageous attempt was made to rectify a major environmental problem that, hitherto, had been kept out of most people's view. Just eat your bacon and leave the mess for someone else to clean-up. We were fighting an international system of ignorance, apathy, and gross negligence (and extreme profits).

This was a system that allowed corporate pollution to be subsidized by the environment and the public health of people, typically poor people of color.[1] Regardless of the state this was and is still true.[2] This is also still true in other countries where the poor and disenfranchised suffer odor, air, water, and public health concerns due to CAFOs.[3] Environmental justice and CAFOs are an anguishing global worry.[4]

Dead pigs, dead fish, and dead water.

In hindsight, the default setting here is a sickening feeling of disgust and disillusionment. Everyone had sacrificed so much. Working at the LSU Swine Unit, sweating alongside hogs, covered in slurry and roaches, and suffering the indignity of a complete failure. Some people lost their money. A lot of money. George Miller lost his car and home, all his money, then his family, and finally his life. He paid the biggest price. He never got the chance to learn from his mistakes. George had a giant heart. He loved like an open field full of Spring. I would have never met George Miller but for pig waste. All the smells, sweat, and disappointments were worth it. His soul was vast, limitless.

It is easy to assume that characters like Blobby, Jim Tryand, D. Reggie, and Shaky Eddie were obvious con-artists, and anyone with common sense would have seen through their lies. It may seem that way to an outsider. Maybe if I had paid more attention I would have seen through the menagerie of lies and the mountain of falsehoods.

Eventually, we all did.

But it was too late.

So, what did I learn? Did I have anything to show for a decade of failure? There were successes. Like wildflowers in a field of thorns and stickers. Some deliberate. Some a bit fortuitous. Many welcomed. Others not so much.

On the professional side, I started a consulting company in 1998, and it still exists today. Not all pirates have ragged wooden ships with the Jolly Roger flying above head. Because of my experience in international environmental affairs, I have helped many clients navigate treacherous waters where swash-bucklers take the form of government bureaucrats or corporations officials. This company has taken me throughout the Middle East, Europe, South and Central America, and Southeast Asia. In addition, I did work in Iowa because of the pig project and the expertise I gained in compost and permitting. I consulted there for two years. Eventually, I was asked to serve in the cabinet of a governor.

So that my wife could spend more time raising the kids, I started a bottled water company. That company supplemented her lost income (she was the breadwinner when we were first married). We ran this company from our spare bedroom and garage. It was very simple: we made up names for companies that wanted their own bottled water. After a while, we created our own label and partnered with convenient stores throughout Louisiana. This experience came in handy when years later I was asked to shepherd a bottled water project from idea to products in the store in northeast Louisiana.[5]

I produced and directed two more environmental film series,[6] and my first stand-alone film about the loss of smart, young people ('brain drain') in Louisiana. That film was shown at festivals across the US.[7] The environmental series were translated into more than 180 languages and have been distributed around the world. I was honored by Romania during Earth Week in 2005. They showed all my films and toured me through their beautiful country. One of my favorite memories was playing ping pong at a resort on the Black Sea. That would never have happened except for this project.

To be fair, all my experiences with the pig project have benefited my career by increasing my knowledge of environmental management, regulations, technology, engineering, and economics.

There were also some personal victories. My son was born. Because of him, I started watching football (and coaching) again. Also, when he was old enough, I hunted for the first time in more than a decade. Later, I purchased a home. Since I never thought that I would get married, have children, and buy a house, all of these were achievements that I never considered likely. If those things were the only successes, then I could say it was all worth it.

The truth, though, remains.

The pig project tossed my ego to the curb and ran me over like roadkill. It remains the biggest failure of my life. I put so many aspects of my career on hold because of it. What did I learn? Never trust anyone that shows up with promises and nothing else. Ask for the budget first. If someone wants to benefit from your hard work, then it is OK to ask for money. If they do not have any, and lack any expertise, then walk away from them quickly. Especially if these are "professional dealmakers" without a regular source of income or a job.

Also, there are buzzards everywhere. Circling overhead. Financial vultures ready to swoop down upon your projects at the slightest hint of vulnerability. Maybe because you need more funding? Or maybe additional expertise? Often, they come as friends or relatives. They always seem to want to "do you a favor" by "helping you out." I have learned to simply keep quiet regarding my work.

Not secretive.

Just private.

This is of great importance when you are struggling. Everyone has a comforter like Eliphaz, Bildad or Zophar in their lives.[8] When they show up, call them out as did Job. Some people relish in the idea of your failure. I try not to give too much ammunition.

For any environmental researcher, I would offer this: do not let the hog wash ruin your project. The seductive lure of money crippled SSA Global. It destroyed George. It delayed my career by a decade. The quest for Morgenthau black boxes filled with Nazi gold, or Vatican wealth blinded us to reality. It made us greedy. Greed is certainly a sin.[9]

But so is pride. We believed our own hog wash and we fell hard. Pride is a close cousin of vanity.

Vanity, all is vanity.[10]

Most things in life are a bad bet. You must make it count and leave this Earth better than how you found it. I try. Sometimes I even succeed.

NOTES

1. Carrel, M., Young, S. G., & Tate, E. (2016). Pigs in space: Determining the environmental justice landscape of swine concentrated animal feeding operations (CAFOs) in Iowa. *International Journal of Environmental Research and Public Health, 13*(9), 849.

2. Nicole, W. (2013). CAFOs and environmental justice: The case of North Carolina. The state seems to be a testing ground for every imaginable environmental justice and race issue.

3. Gladkova, E. (2020). Farming intensification and environmental justice in Northern Ireland. *Critical Criminology, 28*(3), 445–461. An excellent effort that will

surprise many Americans thinking that environmental justice issues only occur when coupled with racial disparities.

4. Davies, P. A. (2014). Green crime and victimization: Tensions between social and environmental justice. *Theoretical Criminology, 18*(3), 300–316.

5. KTVE. King Springs Water bottles spring water in West Monroe straight from the ground. November 16, 2020. Located at https://www.myarklamiss.com/news /king-springs-water-bottles-spring-water-in-west-monroe-straight-from-the-ground /. Accessed 9 July 2022. In another article, the *News Star* reported, "This effort included students, faculty, and staff across many departments, colleges, and majors. My sincere thanks to Dr. John Sutherlin for his efforts in leading and coordinating ULM's participation in this impactful and innovative process," said Berry. "Fulfilling this dream and other dreams is ULM's mission." See King Springs water company opens with help from ULM. November 3, 2020. https://www.thenewsstar.com/story/ news/education/2020/11/03/ulm-helps-develop-king-springs-water-company-in-west -monroe/6130957002/. Accessed 9 July 2022.

6. Reclaiming Our Urban Environment. Located at https://www.films.com/id /6841/Brownfields_Reclaiming_Our_Urban_Environment.htm. Accessed 10 July 2022. This film took me across the US for almost two years with an excellent film crew.

7. Stay Brady Stay. Trailer. Located at https://www.youtube.com/watch?v=_ _C3y37C2lg. Accessed 10 May 2022. I showed this film in San Francisco and then in San Antonio. It was featured in two other festivals. The ability to see my film on a huge screen was overwhelming.

8. Job 16:2. Job's "miserable comforters." Job has so many lessons that it is often difficult to limit to one story.

9. Tickle, P. A. (2004). *Greed: The seven deadly sins*. Oxford University Press. Strangely enough, greed seems to be OK these days. Is there too much of anything anymore? Maybe Gordon Gekko (Wall Street, 1987) made it acceptable?

10. Ecclesiastes 1:2.

Select Bibliography

Aarnink, A. J. A., & Verstegen, M. W. A. (2007). Nutrition, key factor to reduce environmental load from pig production. *Livestock Science, 109*(1–3), 194–203.

Aira, M., Monroy, F., & Domínguez, J. (2007). Earthworms strongly modify microbial biomass and activity triggering enzymatic activities during vermicomposting independently of the application rates of pig slurry. *Science of the Total Environment, 385*(1–3), 252–261.

An, B. X., & Preston, T. R. (1999). Gas production from pig manure fed at different loading rates to polyethylene tubular biodigesters. *Livestock Research for Rural Development, 11*(1), 1–8.

Anderson, N., Strader, R., & Davidson, C. (2003). Airborne reduced nitrogen: ammonia emissions from agriculture and other sources. *Environment International, 29*(2–3), 277–286.

Aneja, V. P., Schlesinger, W. H., Nyogi, D., Jennings, G., Gilliam, W., Knighton, R. E., ... & Krishnan, S. (2006). Emerging national research needs for agricultural air quality. *Eos, Transactions American Geophysical Union, 87*(3), 25–29.

Angelidaki, I., & Ahring, B. K. (2000). Methods for increasing the biogas potential from the recalcitrant organic matter contained in manure. *Water Science and Technology, 41*(3), 189–194.

Ashwood, L. (2012). Daniel Imhoff (Ed): The CAFO reader: the tragedy of industrial animal factories. *Agriculture and Human Values, 29*(3), 427–428.

Barker, J. C., & Zublena, J. P. (1995). *Livestock manure nutrient assessment in North Carolina. Final Report.* Raleigh, NC: North Carolina Agricultural Extension Service, North Carolina State University.

Baylis, K., Peplow, S., Rausser, G., & Simon, L. (2008). Agri-environmental policies in the EU and United States: A comparison. *Ecological Economics, 65*(4), 753–764.

Behr, P., & Schmidt, R. H. (2016). The German banking system. In *The Palgrave handbook of European banking* (pp. 541–566). Palgrave Macmillan, London.

Ben, W., Pan, X., & Qiang, Z. (2013). Occurrence and partition of antibiotics in the liquid and solid phases of swine wastewater from concentrated animal feeding operations in Shandong Province, China. *Environmental Science: Processes & Impacts, 15*(4), 870–875.

Benarde, M. A., Israel, B. M., Olivieri, V. P., & Granstrom, M. L. (1965). Efficiency of chlorine dioxide as a bactericide. *Applied Microbiology, 13*(5), 776–780.

Boinon, J. P., & Hoetjes, B. J. S. (1999). Netherlands: from waste quotas to pig quotas. *L'agriculture européenne et les droits à produire.*, 271–287.

Bonmati, A., Flotats, X., Mateu, L., & Campos, E. (2001). Study of thermal hydrolysis as a pretreatment to mesophilic anaerobic digestion of pig slurry. *Water Science and Technology, 44*(4), 109–116.

Bradford, S. A., Segal, E., Zheng, W., Wang, Q., & Hutchins, S. R. (2008). Reuse of concentrated animal feeding operation wastewater on agricultural lands. *Journal of Environmental Quality, 37*(S5), S-97.

Braunstein, S., & Lavizzo-Mourey, R. (2011). How the health and community development sectors are combining forces to improve health and well-being. *Health Affairs, 30*(11), 2042–2051.

Brooks, J. P., Adeli, A., & McLaughlin, M. R. (2014). Microbial ecology, bacterial pathogens, and antibiotic resistant genes in swine manure wastewater as influenced by three swine management systems. *Water Research, 57*, 96–103.

Buchanan, Kirsty. *The Telegraph*. The Senior Tories with the secret Soviet past. March 19, 2021. Located at https://www.telegraph.co.uk/news/2021/03/19/senior -tories-secret-soviet-past/. Accessed 30 June 2022.

Bujoczek, G., Oleszkiewicz, J., Sparling, R. R. C. S., & Cenkowski, S. (2000). High solid anaerobic digestion of chicken manure. *Journal of Agricultural Engineering Research, 76*(1), 51–60.

Bunton, B., O'Shaughnessy, P., Fitzsimmons, S., Gering, J., Hoff, S., Lyngbye, M., . . . & Werner, M. (2007). Monitoring and modeling of emissions from concentrated animal feeding operations: overview of methods. *Environmental Health Perspectives, 115*(2), 303–307.

Burkholder, J., Libra, B., Weyer, P., Heathcote, S., Kolpin, D., Thorne, P. S., & Wichman, M. (2007). Impacts of waste from concentrated animal feeding operations on water quality. *Environmental Health Perspectives, 115*(2), 308–312.

Bustamante, M. A., Moral, R., Bonmatí, A., Palatsí, J., Solé-Mauri, F., & Bernal, M. P. (2014). Integrated waste management combining anaerobic and aerobic treatment: A case study. *Waste and Biomass Valorization, 5*(3), 481–490.

Carrel, M., Young, S. G., & Tate, E. (2016). Pigs in space: Determining the environmental justice landscape of swine concentrated animal feeding operations (CAFOs) in Iowa. *International Journal of Environmental Research and Public Health, 13*(9), 849.

Chappell, H. W. (1982). Campaign contributions and congressional voting: A simultaneous probit-tobit model. *The Review of Economics and Statistics*, 77–83.

Chastain, J. P., Camberato, J. J., Albrecht, J. E., & Adams, J. (1999). Swine manure production and nutrient content. South Carolina confined animal manure managers certification program. Clemson University, SC, 1–17.

Chatterjee, N., Flury, M., Hinman, C., & Cogger, C. G. (2013). Chemical and physical characteristics of compost leachates. A Review Report prepared for the Washington State Department of Transportation. Washington State University.

Chefetz, B., Hatcher, P. G., Hadar, Y., & Chen, Y. (1996). Chemical and biological characterization of organic matter during composting of municipal solid waste (Vol. 25, No. 4, pp. 776–785). American Society of Agronomy, Crop Science Society of America, and Soil Science Society of America.

Chin, K. K., & Ong, S. L. (1993). A wastewater treatment system for an industrialized pig farm. *Water Science and Technology, 28*(7), 217–222.

Clark, C. E. (1965). Hog waste disposal by lagooning. *Journal of the Sanitary Engineering Division, 91*(6), 27–42.

Cullimore, D. R., Maule, A., & Mansuy, N. (1985). Ambient temperature methanogenesis from pig manure waste lagoons: thermal gradient incubator studies. *Agricultural Wastes, 12*(2), 147–157.

Davies, P. A. (2014). Green crime and victimization: Tensions between social and environmental justice. *Theoretical Criminology, 18*(3), 300–316.

DEFRA. Located at https://www.gov.uk/government/organisations/department-for -environment-food-rural-affairs. Accessed 27 June 2022.

Deshwal, B. R., & Lee, H. K. (2005). Manufacture of chlorine dioxide from sodium chlorite: Process chemistry. *Journal of Industrial and Engineering Chemistry, 11*(1), 125–136.

Doane, M. (2014). Politics and the family farm: When the neighbors poison the well. *Anthropology Now, 6*(3), 45–52.

Donham, K. J., Wing, S., Osterberg, D., Flora, J. L., Hodne, C., Thu, K. M., & Thorne, P. S. (2007). Community health and socioeconomic issues surrounding concentrated animal feeding operations. *Environmental Health Perspectives, 115*(2), 317–320.

Dorca-Preda, T., Mogensen, L., Kristensen, T., & Knudsen, M. T. (2021). Environmental impact of Danish pork at slaughterhouse gate–a life cycle assessment following biological and technological changes over a 10-year period. *Livestock Science*, 251, 104622.

Eniola, B., Perschbacher-Buser, Z., Caraway, E., Ghosh, N., Olsen, M., & Parker, D. (2006). Odor control in waste management lagoons via reduction of p-cresol using horseradish peroxidase. In 2006 ASAE Annual Meeting (p. 1). American Society of Agricultural and Biological Engineers.

Fare, G., & Potts, S. E. (1987). Rising ffederal waters: The nation's quagmire. *Baylor Texas Environmental Law Journal*, 18, 1.

Fine, Ken and Erica Hellerstein. *Indy Week*. Big pork has given $272,000 to House Republicans who voted in favor of hog-farm-protection bill. April 7, 2017. Located at https://indyweek.com/news/archives/big-pork-given-272-000-house-republicans -voted-favor-hog-farm-protection-bill/. Accessed 10 June 2022.

Fisher, K. L., & Hoffmans, L. W. (2009). *How to smell a rat: The five signs of financial fraud*. John Wiley & Sons.

Fluidized Bed Reactor Apparatus. Located at https://patents.google.com/patent/ US20030209476A1/en. Accessed 4 January 2022. The official patent number is US20030209476A1.

Foltz, J., Barham, B., & Kim, K. (2000). Universities and agricultural biotechnology patent production. *Agribusiness: An International Journal, 16*(1), 82–95.

Furuseth, O. J. (1997). Restructuring of hog farming in North Carolina: Explosion and implosion. *The Professional Geographer, 49*(4), 391–403.

Gallert, C., Fund, K., & Winter, J. (2005). Antibiotic resistance of bacteria in raw and biologically treated sewage and in groundwater below leaking sewers. *Applied Microbiology and Biotechnology, 69*(1), 106–112.

Ganglmair, B., Robinson, W. K., & Seeligson, M. (2022). The rise of process claims: Evidence from a century of US patents. ZEW-Centre for European Economic Research Discussion Paper, (22-011).

Gettier, S. W., & Roberts, M. (1994). Swine lagoon biogas utilization system (No. CONF-9410176-). Western Regional Biomass Energy Program, Reno, NV (United States).

Gladkova, E. (2020). Farming intensification and environmental justice in Northern Ireland. *Critical Criminology, 28*(3), 445–461.

Goldberg, A. M. (2016). Farm animal welfare and human health. *Current Environmental Health Reports, 3*(3), 313–321.

Goldblatt, M. E., Eble, K. S., & Feathers, J. E. (1993). Zero discharge: What, why, and how. *Chemical Engineering Progress*;(United States), *89*(4).

Gordon, G., & Rosenblatt, A. A. (2005). Chlorine dioxide: The current state of the art. *Ozone: Science & Engineering, 27*(3), 203–207.

Graham, J. P., & Nachman, K. E. (2010). Managing waste from confined animal feeding operations in the United States: The need for sanitary reform. *Journal of Water and Health, 8*(4), 646–670.

Greger, M., & Koneswaran, G. (2010). The public health impacts of concentrated animal feeding operations on local communities. *Family and Community Health*, 11–20.

Griffin, G. E., & Bramhall, M. (1998). *The creature from Jekyll Island: A second look at the Federal Reserve* (p. 608). American Media.

Griffiths, A. J., & Hicks, W. (1997). Agricultural waste to energy—a UK perspective. *Energy & Environment, 8*(2), 151–167.

Halberg, N., Verschuur, G., & Goodlass, G. (2005). Farm level environmental indicators; are they useful? An overview of green accounting systems for European farms. *Agriculture, Ecosystems & Environment, 105*(1–2), 195–212.

Hammond, E. G., Heppner, C., & Smith, R. (1989). Odors of swine waste lagoons. *Agriculture, Ecosystems & Environment, 25*(2–3), 103–110.

Happe, K. (2004). Agricultural policies and farm structures-agent-based modelling and application to EU-policy reform (No. 920-2016-72838).

Harden, S. L. (2015). Surface-water quality in agricultural watersheds of the North Carolina Coastal Plain associated with concentrated animal feeding operations (No. 2015-5080). US Geological Survey.

Harmin, C. (2015). Flood vulnerability of hog farms in Eastern North Carolina: An inconvenient poop.

Hart, S. A., & Turner, M. E. (1965). Lagoons for livestock manure. *Journal (Water Pollution Control Federation)*, 1578–1596.

Have, H., & Henriksen, K. S. (1998). An energy-efficient combustion system for high-moisture organic wastes and biomasses. *Water and Environment Journal, 12*(3), 224–232.

Head III, T. R. (1999). Local regulation of animal feeding operations: Concerns limits, and options for southeastern states. *Environmental Law, 6*, 503.

Heederik, D., Sigsgaard, T., Thorne, P. S., Kline, J. N., Avery, R., Bønløkke, J. H., . . . & Merchant, J. A. (2007). Health effects of airborne exposures from concentrated animal feeding operations. *Environmental Health Perspectives, 115*(2), 298–302.

Hentz Jr, L. H., Murray, C. M., Thompson, J. L., Gasner, L. L., & Dunson Jr, J. B. (1992). Odor control research at the Montgomery County regional composting facility. *Water Environment Research, 64*(1), 13–18.

Higashikawa, F. S., Silva, C. A., Nunes, C. A., Bettiol, W., & Guerreiro, M. C. (2016). Physico-chemical evaluation of organic wastes compost-based substrates for Eucalyptus seedlings growth. *Communications in Soil Science and Plant Analysis, 47*(5), 581–592.

Hobbs, P. J., Misselbrook, T. H., & Cumby, T. R. (1999). Production and emission of odours and gases from ageing pig waste. *Journal of Agricultural Engineering Research, 72*(3), 291–298.

Hsu, J. H., & Lo, S. L. (1999). Chemical and spectroscopic analysis of organic matter transformations during composting of pig manure. *Environmental Pollution, 104*(2), 189–196.

Huang, G. F., Wu, Q. T., Wong, J. W. C., & Nagar, B. B. (2006). Transformation of organic matter during co-composting of pig manure with sawdust. *Bioresource Technology, 97*(15), 1834–1842.

Hue, N. V., & Sobieszczyk, B. A. (1999). Nutritional values of some biowastes as soil amendments. *Compost Science & Utilization, 7*(1), 34–41.

Hunt, P. G., & Poach, M. E. (2001). State of the art for animal wastewater treatment in constructed wetlands. *Water Science and Technology, 44*(11–12), 19–25.

Indriyati, L. T., & Goto, I. (1997). Effect of zeolite addition to chicken manure on nitrogen mineralization in the soil. In *Plant nutrition for sustainable food production and environment* (pp. 593–594). Springer, Dordrecht.

Irgens, R. L., & Day, D. L. (1966). Laboratory studies of aerobic stabilization of swine waste. *Journal of Agricultural Engineering Research, 11*(1), 1–10.

Jin, R. Y., Hu, S. Q., Zhang, Y. H., & Bo, T. (2008). Research on the explosion characteristics of chlorine dioxide gas. *Chinese Chemical Letters, 19*(11), 1375–1378.

Jongbloed, A. W., & Henkens, C. H. (1996). Environmental concerns of using animal manure—the Dutch case. Nutrient management of food animals to enhance and protect environment. *Lewis Publishers*, 315–332.

Kalyuzhnyi, S., Sklyar, V., Rodriguez-Martinez, J., Archipchenko, I., Barboulina, I., Orlova, O., . . . & Klapwijk, A. (2000). Integrated mechanical, biological and physico-chemical treatment of liquid manure streams. *Water Science and Technology, 41*(12), 175–182.

Kelly-Reif, K., & Wing, S. (2016). Urban-rural exploitation: An underappreciated dimension of environmental injustice. *Journal of Rural Studies, 47*, 350–358.

Koerkamp, P. G., Metz, J. H. M., Uenk, G. H., Phillips, V. R., Holden, M. R., Sneath, R. W., . . . & Wathes, C. M. (1998). Concentrations and emissions of ammonia in livestock buildings in Northern Europe. *Journal of Agricultural Engineering Research, 70*(1), 79–95.

Kornegay, E. T., Harper, A. F., Jones, R. D., & Boyd, L. J. (1997). Environmental nutrition: Nutrient management strategies to reduce nutrient excretion of swine. *The Professional Animal Scientist, 13*(3), 99–111.

Ladd, A. E., & Edward, B. (2002). Corporate swine and capitalist pigs: A decade of environmental injustice and protest in North Carolina. *Social Justice, 29*(3 (89), 26–46.

Laitos, J. G., & Ruckriegle, H. (2012). The clean water act and the challenge of agricultural pollution. *Vermont Law Review, 37,* 1033.

Letson, D., Gollehon, N., Kascak, C., Breneman, V., & Mose, C. (1998). Confined animal production and groundwater protection. *Applied Economic Perspectives and Policy, 20*(2), 348–364.

Liu, X., Zhang, W., Hu, Y., Hu, E., Xie, X., Wang, L., & Cheng, H. (2015). Arsenic pollution of agricultural soils by concentrated animal feeding operations (CAFOs). *Chemosphere, 119,* 273–281.

Loehr, R. C. (1969). Animal wastes—a national problem. *Journal of the Sanitary Engineering Division, 95*(2), 189–222.

Lopez, R. A. (2001). Campaign contributions and agricultural subsidies. *Economics & Politics, 13*(3), 257–279.

Lynch, B. C. (2012). International patent harmonization: Creating a binding prior art search within the patent cooperation treaty. *Geo. Wash. Int'l L. Rev., 44,* 403.

Mallin, M. A., McIver, M. R., Robuck, A. R., & Dickens, A. K. (2015). Industrial swine and poultry production cause chronic nutrient and fecal microbial stream pollution. *Water, Air, & Soil Pollution, 226*(12), 1–13.

Marr, J. B., & Facey, R. M. (1995). Agricultural waste. *Water Environment Research, 67*(4), 503–507.

Martin Jr, J. H. (1997). The clean water act and animal agriculture. *Journal of Environmental Quality, 26*(5), 1198–1203.

McGinley, M. A., & McGinley, C. M. (2001). The new European olfactometry standard: implementation, experience, and perspectives. In Air and Waste Management Association, Annual Conference Technical Program, Session No. EE-6b: Modelling, Analysis and Management of Odours.

Miner, J. R., Goh, A. C., & Taiganides, E. P. (1983). Dewatering anaerobic swine manure lagoon sludge using a decanter centrifuge. *Transactions of the ASAE, 26*(5), 1486–1489.

Mirabelli, M. C., Wing, S., Marshall, S. W., & Wilcosky, T. C. (2006). Race, poverty, and potential exposure of middle-school students to air emissions from confined swine feeding operations. *Environmental Health Perspectives, 114*(4), 591–596.

Mitloehner, F. M., & Calvo, M. S. (2008). Worker health and safety in concentrated animal feeding operations. *Journal of Agricultural Safety and Health, 14*(2), 163–187.

Mkhabela, M. S., Gordon, R., Burton, D., Smith, E., & Madani, A. (2009). The impact of management practices and meteorological conditions on ammonia and nitrous oxide emissions following application of hog slurry to forage grass in Nova Scotia. *Agriculture, Ecosystems & Environment, 130*(1–2), 41–49.

Mohaibes, M., & Heinonen-Tanski, H. (2004). Aerobic thermophilic treatment of farm slurry and food wastes. *Bioresource Technology, 95*(3), 245–254.

Moore, R., & Barnes, J. (2004). Faces from the flood: Hurricane Floyd remembered. UNC Press Books. The authors donated all proceeds from the book to the American Red Cross.

Moser, M. A. (1998). Anaerobic digesters control odors, reduce pathogens, improve nutrient manageability, can be cost competitive with lagoons, and provide energy too. Resource Conservation Management, Inc. Presentation at Iowa State University.

Mounsey, R. J., & Hagar, M. C. (1946). Taste and odor control with chlorine dioxide. *Journal (American Water Works Association), 38*(9), 1051–1056.

Mussari, F. P., Smith, J. E., & Jr, D. M. (2013). Accelerated composting methods and equipment. *European Journal of Marketing.*

Nazir, M. (1991). Biogas plants construction technology for rural areas. *Bioresource Technology, 35*(3), 283–289.

Ndegwa, P. M., Hristov, A. N., Arogo, J., & Sheffield, R. E. (2008). A review of ammonia emission mitigation techniques for concentrated animal feeding operations. *Biosystems Engineering, 100*(4), 453–469.

Neel, J. K., & Hopkins, G. J. (1956). Experimental lagooning of raw sewage. *Sewage and Industrial Wastes, 28*(11), 1326–1356.

Nguyen, T. L. T., Hermansen, J. E., & Mogensen, L. (2012). Environmental costs of meat production: The case of typical EU pork production. *Journal of Cleaner Production, 28*, 168–176.

Nicole, W. (2013). CAFOs and environmental justice: The case of North Carolina.

Ogishi, A., Metcalfe, M. R., & Zilberman, D. (2002). Animal waste policy: reforms to improve environmental quality. *Choices, 17*(316-2016-7143).

Oglesby, Cameron. Environmental Health News. Hurricane season spurs hog waste worries in North Carolina. Located at https://www.ehn.org/north-carolina-hurricanes-hog-farms-2652972415/hog-farms-in-north-carolinas-100-year-flood-plain. Accessed January 13, 2022.

Pain, B. F., & Klarenbeek, J. V. (1988). Anglo-Dutch experiments on odour and ammonia emissions from landspreading livestock wastes (No. 88-2). IMAG.

Parker, D. B., Caraway, E. A., Rhoades, M. B., Cole, N. A., Todd, R. W., & Casey, K. D. (2010). Effect of wind tunnel air velocity on VOC flux from standard solutions and CAFO manure/wastewater. *Transactions of the ASABE, 53*(3), 831–845.

Paul Taiganides, E. (1986). Animal farming effluent problems–an integrated approach: Resource recovery in large scale pig farming. *Water Science and Technology, 18*(3), 47–55.

Peters, K. A. (2010). Creating a sustainable urban agriculture revolution. *Journal of Environmental Law and Litigation, 25*, 203.

Persson, S. P., Bartlett, H. D., Branding, A. E., & Regan, R. W. (1979). Agricultural anaerobic digesters: Design and operation (No. NP-2901472). Pennsylvania State Univ., University Park (USA). Agricultural Experiment Station.

Pew Commission on Industrial Farm Animal Production. Located at https://www.pcifapia.org/. Accessed 2 June 2022.

Possas, M. L., Salles-Filho, S., & da Silveira, J. (1996). An evolutionary approach to technological innovation in agriculture: some preliminary remarks. *Research Policy, 25*(6), 933–945.

Potter, C., & Lobley, M. (1993). Helping small farms and keeping Europe beautiful: a critical review of the environmental case for supporting the small family farm. *Land Use Policy, 10*(4), 267–279.

Pratt, S. J., Frarey, L., & Carr, A. (1997). A comparison of US and UK law regarding pollution from agricultural runoff. *Drake L. Rev., 45*, 159.

Price, C. (2015). *Odour regulations in Europe—different approaches.* CERC Publication.

Pruden, A. (2009). *Hormones and pharmaceuticals generated by concentrated animal feeding operations.* L. S. Shore (Ed.). Springer New York.

Prummel, J. (1960). Placement of a compound (NPK) fertilizer compared with straight fertilizers. *Netherlands Journal of Agricultural Science, 8*(2), 149–154.

Pytlar Jr, T. S. (2010, January). Status of existing biomass gasification and pyrolysis facilities in North America. In North American Waste-to-Energy Conference (Vol. 43932, pp. 141–154).

Radelyuk, I., Tussupova, K., & Zhapargazinova, K. (2019). Assessment of groundwater safety around contaminated water storage sites. In *11th world congress on water resources and environment: Managing water resources for a sustainable future-EWRA 2019.* Proceedings.

Ram, N. (2009). Assigning rights and protecting interests: Constructing ethical and efficient legal rights in human tissue research. *Harvard Journal of Law and Technology, 23*, 119.

Raven, R., & Verbong, G. (2004). Dung, sludge, and landfill: Biogas technology in the Netherlands, 1970–2000. *Technology and Culture, 45*(3), 519–539.

Roberts, R. G., & Reymond, S. T. (1994). Chlorine dioxide for reduction of postharvest pathogen inoculum during handling of tree fruits. *Applied and Environmental Microbiology, 60*(8), 2864–2868.

Röling, N. G. (1988). *Extension science: Information systems in agricultural development.* Cambridge University Press.

Romero, C., Ramos, P., Costa, C., & Márquez, M. C. (2013). Raw and digested municipal waste compost leachate as potential fertilizer: comparison with a commercial fertilizer. *Journal of Cleaner Production, 59*, 73–78.

Rowan, J. P., Durrance, K. L., Combs, G. E., & Fisher, L. Z. (1997). The digestive tract of the pig. Animal Science Department, Florida Cooperative Extension Service, Institute of Food and Agricultural Sciences, University of Florida, Gainesville, Document AS23, 1(4), 1–7.

Ros, M., Garcia, C., & Hernández, T. (2006). A full-scale study of treatment of pig slurry by composting: Kinetic changes in chemical and microbial properties. *Waste Management, 26*(10), 1108–1118.

Rosov, K. A., Mallin, M. A., & Cahoon, L. B. (2020). Waste nutrients from US animal feeding operations: Regulations are inconsistent across states and inadequately assess nutrient export risk. *Journal of Environmental Management, 269*, 110738.

Ruhl, J. B. (2000). Farms, their environmental harms, and environmental law. *Ecology LQ, 27*, 263.

Saeys, W., Mouazen, A. M., & Ramon, H. (2005). Potential for onsite and online analysis of pig manure using visible and near infrared reflectance spectroscopy. *Biosystems Engineering, 91*(4), 393–402.

Safley Jr, L. M., & Westerman, P. W. (1988). Biogas production from anaerobic lagoons. *Biological Wastes, 23*(3), 181–193.

Sambo, A. S., Garba, B., & Danshehu, B. G. (1995). Effect of some operating parameters on biogas production rate. *Renewable Energy, 6*(3), 343–344.

Sato, A. (2005). Public participation andaccess to clean water: An analysis of the CAFO rule. *Sustainable Development and Law and Policy, 5*, 40.

Schmidt, C. W. (2000). Lessons from the flood: Will Floyd change livestock farming? *Environmental Health Perspectives, 108*(2), A74–A77.

Schultz, A. A., Peppard, P., Gangnon, R. E., & Malecki, K. M. (2019). Residential proximity to concentrated animal feeding operations and allergic and respiratory disease. *Environment International, 130*, 104911.

Seidl, A., & Grannis, J. (1998). Swine policy decision points. Agricultural and resource policy report (Colorado State University. Dept. of Agricultural and Resource Economics); ARPR 98–01.

Sharpe, R. R., & Harper, L. A. (1999). Methane emissions from an anaerobic swine lagoon. *Atmospheric Environment, 33*(22), 3627–3633.

Sousan, S., Iverson, G., Humphrey, C., Lewis, A., Streuber, D., & Richardson, L. (2021). High-frequency assessment of air and water quality at a concentration animal feeding operation during wastewater application to spray fields. *Environmental Pollution, 288*, 117801.

Spellman, F. R., & Whiting, N. E. (2007). Environmental management of concentrated animal feeding operations (CAFOs). CRC Press.

Stoltenberg, D. H., & McKinney, R. E. (1966). Discussion of "hog waste disposal by lagooning." *Journal of the Sanitary Engineering Division, 92*(4), 78–81.

Stolze, M., Piorr, A., Häring, A. M., & Dabbert, S. (2000). *Environmental impacts of organic farming in Europe.* Universität Hohenheim, Stuttgart-Hohenheim.

Stratmann, T. (1991). What do campaign contributions buy? Deciphering causal effects of money and votes. *Southern Economic Journal*, 606–620.

Sutherland, E. J. (1999). *The siting of concentrated animal feeding operations (CAFOs): Information gaps for achieving environmental justice.* Georgia Institute of Technology.

Tabachow, R. M., Peirce, J. J., & Essiger, C. (2001). Hurricane-Loaded Soil: Effects on Nitric Oxide Emissions from Soil. *Journal of Environmental Quality, 30*(6), 1904–1910.

Taylor, D. A. (2001). From pigsties to hog heaven? *Environmental Health Perspectives, 109*(7), A328–A331.

Thompson, C. D., & Amberg, R. (2001). The great deluge: A chronicle of the aftermath of Hurricane Floyd. *Southern Cultures, 7*(3), 65–82.

Tilche, A., & Bortone, G. (1992). Biological purification and recovery of biogas from pig slurry. *Informatore Agrario, 48*(18 (Supplement)), 51–55.

Tiquia, S. M., Tam, N. F. Y., & Hodgkiss, I. J. (1998). Changes in chemical properties during composting of spent pig litter at different moisture contents. *Agriculture, Ecosystems & Environment, 67*(1), 79–89.

Tremblay, K. R., & Dunlap, R. E. (1978). Rural-urban residence and concern with environmental quality: A replication and extension. *Rural Sociology, 43*(3), 474.

Trusts, P. C., & Hopkins, J. (2008). Putting meat on the table: Industrial farm animal production in America. A Report of the Pew commission on industrial Farm Animal Production.

Turner, K. (2000). A review of US patents in the field of organic process development published during April to July 2000.

Ulrich-Schad, J. D., Babin, N., Ma, Z., & Prokopy, L. S. (2016). Out-of-state, out of mind? Non-operating farmland owners and conservation decision making. *Land Use Policy, 54*, 602–613.

US General Accounting Office. Livestock agriculture: Increased EPA oversight will improve environmental program for concentrated animal feeding operations. January 2003, GAO-03-285, p. 7.

US Government Accountability Office. Concentrated animal feeding operations: EPA needs more information and a clearly defined strategy to protect air and water quality from pollutants of concern. Located at https://www.gao.gov/products/gao-08-944. Accessed 6 June 2022.

US Patent and Trademark Office (USPTO). Located at https://www.uspto.gov/web/offices/pac/mpep/s1824.html. Accessed 23 April 2022.

Van Dyne, D. L., & Weber, J. A. (1994). Biogas production from animal manures: What is the potential? Industrial Uses/IUS-4/Special Article.

Van Harreveld, A. P., Heeres, P., & Harssema, H. (1999). A review of 20 years of standardization of odor concentration measurement by dynamic olfactometry in Europe. *Journal of the Air & Waste Management Association, 49*(6), 705–715.

Van Overwalle, G. (1997). The legal protection of biotechnological inventions in Europe and in the United States: Current framework and future developments.

Walker, D. I. (1999). Rethinking rights of first refusal. *Stanford Journal of Business Law and Finance, 5*, 1.

Walton, L., & Jaiven, K. K. (2020). Regulating CAFOs for the well-being of farm animals, consumers, and the environment. *Environmental Law Report, 50*, 10485.

Webb, J., Menzi, H., Pain, B. F., Misselbrook, T. H., Dämmgen, U., Hendriks, H., & Döhler, H. (2005). Managing ammonia emissions from livestock production in Europe. *Environmental Pollution, 135*(3), 399–406.

Weiland, P. (2010). Biogas production: Current state and perspectives. *Applied Microbiology and Biotechnology, 85*(4), 849–860.

Wen, G., Xu, X., Huang, T., Zhu, H., & Ma, J. (2017). Inactivation of three genera of dominant fungal spores in groundwater using chlorine dioxide: Effectiveness, influencing factors, and mechanisms. *Water Research, 125*, 132–140.

Wielinga, E. (2000). Rural extension in vital networks, changing roles of extension in Dutch agriculture. *Journal of International Agricultural and Extension Education, 7*(1), 23–36.

Windhorst, H. W. (1998). Sectoral and regional patterns of pig production in the EU. *Pig News and Information, 19*(1).

Wong, K. W. (1986). Use of ozone in the treatment of water for potable purposes. *Water Science and Technology, 18*(3), 95.

Wright, P., Inglis, S., Ma, J., Gooch, C., Aldrich, B., Meister, A., & Scott, N. (2004). Comparison of five anaerobic digestion systems on dairy farms. In 2004 ASAE Annual Meeting (p. 1). American Society of Agricultural and Biological Engineers.

Wrigley, T. J., Webb, K. M., & Venkitachalm, H. (1992). A laboratory study of struvite precipitation after anaerobic digestion of piggery wastes. *Bioresource Technology, 41*(2), 117–121.

Wu, J. J. J. (1998). *The use of ozone for the removal of odor from swine manure.* Michigan State University.

Yang, Z., Han, L., & Fan, X. (2006). Rapidly estimating nutrient contents of fattening pig manure from floor scrapings by near infrared reflectance spectroscopy. *Journal of Near Infrared Spectroscopy, 14*(4), 261–268.

Zande, K. (2008). Raising a stink: Why Michigan CAFO regulations fail to protect the state's air and Great Lakes and are in need or revision. *Buffalo Environmental Law Journal, 16*, 1.

Zhang, C., Du, Y., Tao, X. Q., Zhang, K., Shen, D. S., & Long, Y. Y. (2013). Dechlorination of polychlorinated biphenyl-contaminated soil via anaerobic composting with pig manure. *Journal of Hazardous Materials, 261*, 826–832.

Zhang, R., Yamamoto, T., & Bundy, D. S. (1996). Control of ammonia and odors in animal houses by a ferroelectric plasma reactor. *IEEE Transactions on Industry Applications, 32*(1), 113–117.

Index

Page references for figures are italicized.

About the Author

John W. Sutherlin, PhD, is a professor of political dcience and public administration and the Chief Innovation & Research Officer (CIRO) for the University of Louisiana Monroe (ULM). Dr. Sutherlin obtained his PhD from the University of New Orleans (1998) and a master of environmental law and policy from Vermont Environmental Law School (2017). He completed leadership training at the Kennedy School of Government at Harvard (2014).

Dr. Sutherlin was nominated to be secretary of the Department of Environmental Quality (DEQ) for the state of Louisiana under Governor Kathleen Blanco and remains involved in environmental policy, education, and reform in Louisiana. He conducts environmental research and advises governments and corporations in Asia, Europe, the Middle East, and South America. He is a patented inventor, author, and producer of 28 environmental films.

Dr. Sutherlin coauthored *Playing with Fire: The Strange Case of Marine Shale Processors* (2021) with Daniel Elliot Gonzalez.

His second book in this "ecological trilogy" was *Garbage Gumbo: The Strange Tale of the Belle Landfill* (2022).

Made in the USA
Middletown, DE
27 December 2023

46868442R00113